The Green Entrepreneur

Business Opportunities That Can Save the Earth and Make You Money

Gustav Berle

LIBERTY HALL
PRESS™

Save the world.
We might need it later.

Thank you, Hallmark Cards

Remarks of an ecomaniac—

"*I watched green spaces turn into malls, the smell of orange blossoms turn into exhaust fumes . . . we have to ask ourselves, are we the beneficiaries of progress or the victims? There's no point in being asked to read my lips if the lips are not saying anything. That's called lip service . . . How do we best dispose of nuclear waste? By not creating any . . . I came because I care about the environment . . . It may sound square, and I guess I'm old-fashioned but I want to put something back into my own society—and right now, it needs all the help it can get.*"

Robert Redford
The National Press Club
Washington, DC, October 1, 1990

To Esther, who provided the environment
in which this book was able to grow.

FIRST EDITION
FIRST PRINTING

Library of Congress Cataloging-in-Publication Data

Berle, Gustav, 1920-
The green entrepreneur : business opportunities that can save the earth and make you money / by Gustav Berle.
p. cm.
Includes bibliographical references and index.
ISBN 0-8306-0600-9
1. Industrial management—Environmental aspects—United States. 2. Factory and trade waste—Environmental aspects—United States—Management. 3. Pollution—Economic aspects—United States. 4. Commercial products—Environmental aspects—United States. 5. Renewable energy sources—United States. 6. Recycling (Waste, etc.)—United States. 7. Pollution control industry—United States. I. Title.
HD69.P6B47 1991 91-6954
658.4'08—dc20 CIP

For information about other McGraw-Hill materials, call 1-800-2-MCGRAW in the U.S. In other countries call your nearest McGraw-Hill office.

Vice President and Editorial Director: David J. Conti
Book Editor: Lori Flaherty
Production: Katherine G. Brown
Book Design: Jaclyn J. Boone

The entire paperback edition of this book is printed on recycled paper, as are the text portion and book jacket of the hardcover edition. Unfortunately, recycled stock of the required strength was not available for the hardbound cover and endsheets. Perhaps it will be soon.

LHP3

Contents

Preface

ENVIRONMENTALLY RELATED BUSINESSES WILL BE THE HOT GROWTH industry of the 1990s. National expenditures for compliance with federal environmental regulations will pump nearly $100 billion alone into our economy each year. As the revised Clean Air Act takes effect and the multibillion dollar cleanups of hazardous waste sites and government nuclear facilities go forward, the economic importance of these investments will increase even more.

In the decade ahead, our industrial processes must become more efficient and our use of energy more controlled if we are to remain internationally competitive. Materials and by-products will need to be recycled instead of thrown away and new solutions for old problems put into action.

These are exciting times for American entrepreneurs who see the opportunities for greening both here and abroad. *The Green Entrepreneur: Business Opportunities That Can Save the Earth and Make You Money* is the early bird that can help you catch some mighty profitable worms.

Introduction

DO OIL AND WATER MIX? CAN THE QUEST FOR BUSINESS PROFITS BLEND with the concerns for our environment? Many traditionalists have expressed grave doubts that anything short of a cataclysmic catastrophe could bring the two factions together. ''How can the Sierra Club and Greenpeace sleep in the same bed as entrepreneurship?'' is the way one Doubting Thomas put it.

Still, it is possible, and it has been done—successfully. A few years ago, no one would have bet on Mobile Oil Company, AT&T, First Brands (Glad), Waste Management, and even the Chemical Manufacturers Association, to invest thousands of dollars in promoting environmental concerns.

There are banks and investment companies that have not waited for environmentalist pressure to divest themselves of polluters' stocks—they have focused exclusively on investments in companies that clean up their own acts. All over the land, small entrepreneurs are heeding the cry of fouled streams and bays, smog-laden air, and the shrinking ozone layer. These companies use only materials that are ecologically sound and, not surprisingly, an increasingly aware public, supports their efforts by buying their products, even if they cost a few extra pennies.

When consumers talk, big business listens

The "greening" of America, including Canada, is moving forward simply because entrepreneurs can smell opportunities in doing environmentally friendly business. Take a super example, for example, not your typical Vermont organic farmer selling produce to the natural food store in Boston, but an engineering colossus in Fairfax, Virginia. This company was a fairly prosperous engineering firm in the seventies. In the eighties, it saw opportunities in the polluted environment and responded by launching a division that helped the federal government and major industrial firms clean up hazardous waste sites by constructing or reconstructing plants to minimize pollution.

It assessed sites that would be compatible with environmental laws and with the people who lived nearby and offered instruction on making environments safe for labor and how to treat plant waste using natural flow-offs without polluting them. It conducted detailed environmental and geological examinations and, working with regulatory authorities, worked to reduce pollution and contamination and improve water and soil structures through environmental permits and community outreach. They also turned a very nice profit.

The revenue for the firm rose from a few million to $500 million in 1990, and profits rose from a few thousand dollars to more than $8.5 million. In 1989, the company went public and raised some serious money. In the six months during the first half of 1990, the company's stock rose from a little more than $8 a share to nearly $14 a share. In the march to major success, a total of 15 other companies were acquired and a star-studded, politically savvy board of directors was attracted to ensure future fund-raising and business development.

While the torrid pace of this company's growth might not be typical, it illustrates that opportunities abound for entrepreneurs who take their environmentalism seriously.

Another large company, Procter & Gamble, also saw a material advantage in caring for the environment. Since 1986, they have been able to make Pampers and Luvs disposable diapers more absorbent—using half of the cellulose material previously used. And when you do about $1.8 billion of disposable diaper business a year, this isn't peanuts.

Upcoming innovations from the Procter & Gamble laboratories are concentrates that can be mixed with water at home, or super-concentrated soap products, and packaging that is made of lightweight, biodegradable plastic sheeting rather than heavy, bulky cardboard. And you can be sure that if Procter & Gamble is working on these problems, other heavyweights such as Lever Brothers and Colgate-Palmolive are right behind.

In Canada, the Loblaw supermarket chain invested $10 million to launch their own line of environmentally friendly products. More than

100 products carry their new "President's Choice" label. The first few weeks' success prompted one of the chain's executives to state, "Good environmental practice is going to mean good business."

David Nichol, president of Loblaw International Merchants subsidiary, put it this way in a 1989 speech: "Consumers are beginning to comprehend the power that they have to vote for the environment at the cash register. The implications for business are potentially enormous. In the interest of the environment, they are poised to bestow rewards upon your company, or inflict devastating penalties."

Thomas Dahlen, General Manager for Big Bear Markets in San Diego agrees, "Ultimately though, the decision is still the customer's. They vote with their dollars and no one is sure how much they will sacrifice convenience for a product that is more environmentally sound." Big Bear Markets stock more environmentally approved products. They have been working with the Environmental Health Coalition, a private group that evaluates products for toxicity and environmental considerations. By Spring 1990, more than 5 percent of all Big Bear products were thus "approved."

When all is said and done, in a free marketplace, the consumer and the competition determine what course a manufacturer or retailer can take—almost beyond his own inclination to help the environment. This has not been missed on big advertisers, who often bend over backwards to prove their sincerity. Take, for example, this Procter & Gamble advertisement that appeared in a national newspaper, the Boston Globe, in 1989:

> Why We Don't Call Our Diapers Biodegradable
>
> Almost nothing biodegrades in a landfill. That includes every diaper currently on the shelf. Even the ones calling themselves biodegradable.

What is the retail reaction to such candidness? States Big Bear's Tom Dahlen, "Sure we sell cloth diapers, but we still have to sell the items people want to buy. The purpose of *our* program is to stress that there are alternatives."

Small companies heed the call

When it comes to cleaning up the environment—or preventing its despoiling—many small companies are taking the lead. A small but national wood stains and finishes manufacturer, Wood-Kote Products of Portland, Oregon, invested $250,000 hard-earned dollars in building 10, 10,000-gallon, above-ground tanks. These replaced the underground ones that were prone to deteriorating, leaking, and contaminating groundwaters. Stated Walter Thoulion, "We realized that if we were going to be a player in this (environmental) game, if we were going to stay here, somewhere down the line we had to make the commitment."

Douglas Morris, a Washington, DC entrepreneur, had been looking for a business that could provide him a living as well as satisfy his concern for the environment. After nine years in various jobs, he found it. He started a desktop publishing company that published a journal on internship opportunities in the environmental field as well as products that would benefit the environment.

One of these products is a wooden framework that serves as a "bindery" for old newspapers (dovetailing with the demands of the District's new paper recycling law). The first four months, Morris sold 100 of his "Old News Binder" contraptions, operating part-time out of his bedroom on an initial investment of just $2,000 of his own savings.

An automotive repair shop might not sound like a likely prospect for an ecology-minded entrepreneur, but Jeff Shumway, a died-in-the-wool environmentalist and auto mechanic for 13 years, thought there could be a connection. In 1990, he opened Ecotech Autoworks in Tyson's Corner, a Washington suburb in nearby Virginia. Using several thousand dollars of his own savings, he bought an old auto repair shop and, for an additional $6,000, he acquired two machines that recycle Freon and antifreeze. After just four months of operation, Shumway estimated he had nearly halved the amount of garbage formerly generated in his old shop. Today, Shumway's business is so brisk that he plans to double his current staff and franchise his environmentally friendly Ecotech Autoworks repair shop.

There are hundreds of other examples of how small businesses, perhaps armed only with their resolve, ingenuity, and conscience, can blend the profit motive with environmental concerns.

Capitalizing on today's opportunities

The American press has been a major source of bringing the environmental movement to the forefront, pressing the public to become more conscious of, and receptive to, environmentally friendly products and services. In fact, major national publications like *USA Today, The Wall Street Journal, Discover, Fortune, Insight, INC., D&B Reports, The Economist,* and many others have featured articles and whole sections on environmental concerns.

The two-decades-old transition from a lunatic fringe issue to mainstream marketing concerns are making pro-environment efforts this decade's fair-haired bill of rights. Ray A. Goldberg, a Harvard University business professor, commented, "This is not just a small market niche of people who believe in the 'greening of America,' it is becoming a major segment of the consuming public."

The *Green Entrepreneur* can show entrepreneurs how to:

1. Develop manufacturing processes that are nonpolluting and environmentally sound.

2. Develop safe recycling procedures of products that are not being dumped into landfills, endangering people and property.
3. Educate employees and customers to be conscious of the environment in which they and their children will have to live.

You'll examine case histories of entrepreneurs who developed green opportunities to gain more "green," benefitting the earth while enjoying the profitable fruits of their own enterprise. Oil and water can mix—to everyone's advantage.

1

The polluters: big and small, five billion in all

IT IS EASY TO POINT A FINGER AT THE POLLUTERS OF OUR EARTH. AS Pogo said in his famous pronouncement in the comic strip, "We have met the enemy—and it is us." There are no large nations, such as the United States, Russia, India, Brazil, China, and Canada, where pollution is not a problem. In many nations, the accommodation of gigantic, continuously growing populations makes pollution control a frustrating and nearly self-defeating exercise. (Two exceptions are Singapore and Switzerland, from which we can learn a great deal.)

Easier than blaming ourselves and our 5 billion fellow earthlings is to point a finger at the mass polluters—major industries and the 185 million registered automotive vehicles in the United States. Among the industries we must include in this list is agriculture. The runoff from more than 1 billion acres of farmland, much of it sated with pesticides, adds pollution to our water resources.

Big business

Although major industries are not the only culprit, they do contribute a significant amount of pollution to the environment. The environmental scorecard produced by the Boston-based Franklin Research & Development Co. for *Fortune* magazine early in 1990 includes some prominent companies. The following companies were named, and the reasons given, for still needing much improvement in their ecological efforts:

Chemicals	W.R. Grace Co.: toxic dumps and several law suits Monsanto: pesticides, toxic dumps, but good clean-air efforts
Computers	IBM: still indicates high CFC emissions
Oil	Exxon: possibly a fallout from the *Valdez* oil spill Mobil: some polluted sites still found
Photo material	Eastman Kodak: substantial leaks at Rochester site
Steel	Bethlehem Steel: old mills with inherent pollution problems
Environmental services	Browning Ferris: numerous landfill violations Waste Management: steps in progress to reduce environmental hazards

Other companies include General Motors, General Electric, and Borden Co. The vastness and diversification of these companies also lend themselves to violations in hazardous landfills, toxic dumps, air and water complaints, and PCB (polychlorinated biphenyl) cleanup problems.

Large companies that have many divisions will sometimes do environmental good in one section only to have another division undo it. Similarly, an environmentally friendly edict might go out from the boardroom or the public relations department but be ignored by some middle manager down the executive ladder. Companies can also be at fault when one of their product divisions does a marvelous, ecologically proper job but another product division is despoiling the environment. It has happened to the Mobil Oil Company and it has happened to Exxon.

Big brother

Not all big polluters are corporations with smoke-belching chimneys and nonbiodegradable packaging. Many government agencies and municipalities add mind-boggling amounts of pollutants to the atmosphere.

The political face of pollution is sometimes a two-faced one. President Bush came into office as the "environmental president." Some environmentalists think this is a joke. His chief-of-staff, John H. Sununu, has especially been singled out as being anything but environmentally friendly.

During the 1990 Earth Day celebrations, the Sierra Club publicized a list of federal agencies that had the worst record of compliance with the District of Columbia's tough recycling law. Topping the list of scofflaws on what the Sierra Club called the "Slimy Seven," was the Defense Department, which received a "bottom-of-the-trash-barrel award." There followed the White House—which claimed immunity from the D.C. law—and the Department of State, the Small Business Administration (which is looking into recycling plans), the National Credit Union Administration, and the Department of Justice. Ironically, the FBI had an environmental display in its lobby but no working recycling plan. Singled out for praise was the Environmental Protection Agency, which recycles 75 percent of its paper waste, and the Department of Labor, which recycles 64 tons of paper a month.

One of the problems with compliance with its own law is the District's lagging recycling plan. Small private haulers who pick up waste papers and cans for recycling claim that the drop-off and buy-back centers the recycling law was supposed to create have not yet become operative. Evidently, there is room for more entrepreneurs, but the first thing they need to collect is some of the municipal red tape polluting the program.

The biggest villains polluting our cities are not the factories and sewage plants, however, but the routine activities of ever-growing metropolitan areas. Gigantic parking lots at suburban transit depots, shopping malls, arenas, and area schools cause runoffs of oily water each time it rains. Replacing suburban developments, with their acres of parking lots, driveways, and roads where meadows and forests used to stand makes for inalterable changes in our environment. Fertilized lawns, construction site dumps, storm drains, millions of water faucets, air conditioners, refrigerators, barbecues, and other trappings of a burgeoning civilization, create bigger environmental problems than the factories. Take, for instance, what happened in Los Angeles in 1988.

Big cities

For 176 days, the city of Los Angeles and the surrounding area of Southern California violated federal health standards for ozone preservation. It was the worst record for any metropolitan area in the United States. This was no single factory or group that created this health hazard, but a conglomeration of all the businesses in the area and the horrendous automotive traffic. All of which added up to real danger for millions of people.

Los Angeles has 20 years to straighten out its pollution act. A monumental job that will include such drastic measures as phasing out all fossil-fuel-burning vehicles in favor of clean-combusting electric vehicles or those burning nonpolluting alternative fuels. It would mean new and improved, nonsmoking backyard barbecues and electric or other nongaseous-burning lawn mowers and generators. No more charcoal lighter fuel will be sold, nor fluorocarbon spray chemicals. Businesses and factories that are now spewing exhausts into the air will have to have emissions controls installed. There will be group riding, elimination of free and easy parking in the city, staggered work hours, more housing near large employers—anything that will reduce big-scale pollution. The alternative? Evacuate three-quarters of the Los Angeles population into the countryside and install billions of dollars worth of electric commuter rail vehicles.

The *New York Times* commented wrily on the Los Angeles situation: "The plan would be costly. It would also be intrusive . . . If it weren't . . . [if it] didn't challenge cherished habits and sacred cows, it might create some passing wind—and do nothing about the air."

Burgeoning humanity

Demographers, social scientists, and others concerned with the population growth have talked about our "population bomb" going off in our midst . . . that its fuse was burning fast, that its destructive force would forever alter life on earth. What they, and Professor Paul R. Ehrlich, are talking about is the continuing people pollution. While the birthrate in the United States is not far above zero population growth, and many other highly industrialized countries have similar standards, population growth in India, and some nations of Africa continues on a rampant path.

In just the past two decades, the population load of our planet has increased by more than 50 percent—from 3.5 billion to 5.3 billion. While the numbers alone are not frightening, the implications of continued population crowding and environmental pollution, deforestation, land erosion, epidemic possibilities, and starvation are very real indeed.

The grim indicators that come out of the slums of Calcutta, the wastelands of Eritrea and Somalia, the barrios of Mexico City and Rio de Janeiro, the jungles of the Sudan, and the squalid camps of the Palestinians indicate a greatly declining quality of life for millions of earth's inhabitants.

Crowding even in "civilized" and developing countries takes its toll on the environment, and the effects are like the concentric rings of a stone tossed into a still pond. Every day, 216,000 acres of tropical forest are razed to make room for more people and provide wood for building, furniture, and cooking. Every day, each inhabitant generates between 1,000 and 2,000 pounds of toxic wastes (in the United States, it's closer to 2,000 pounds). Each U.S. car emits 60 tons of polluting gas into the atmosphere during its lifetime, and that figure is much more for countries with poorer fuel and lower emission standards.

The statistics of human "pollution" are even more sordid. According to the best estimates available from demographers at the United Nations, drawing from reports from Third World countries, 40,000 infants perish every day from hunger, thirst, disease, and ignorance. The United States is not an innocent bystander either. It has the highest teen pregnancy rate in the world. By the time our children start collecting social security, another 117 million people will have been added to our population, projecting our present rate of growth. With more than 350 million Americans between the Atlantic and the Pacific, the U.S. will have to accommodate 87 new cities the size of Boston or enlarge existing overcrowded cities beyond tolerance.

These figures, and the potential for disaster, are not figments of someone's imagination, but projections by the U.S. Census Bureau. While such growth might tempt entrepreneurs to rub their hands with glee and make anti-abortion zealots jump for joy, the environmental consequences can reduce the quality of life for everyone, of whichever persuasion.

Yet, overpopulation is the best documented, most scientifically controllable environmental problem we have. It is easy to point to ignorance in the Third World where one-half of the globe's population has no, or at best, inadequate, recourse to contraception, but the problem starts right here at home.

(In the five minutes that it has taken you to read this small section on *The most sensitive pollution of all*, 1,200 babies were born and have been added to the earth's population. Truly, it is a human *race*!)

The "environmentally friendly" controversy

As we are at the beginning of the nineties, we find eco-entrepreneuring surrounded by a great deal of confusion. Sure, the public wants to help clean up our environment. In fact, 39 percent of those questioned in a 1990 survey said that they would not mind paying a little more if the products were assuredly environmentally friendly. While only 31 percent of this group could be considered baby boomers, it must be said, in all honesty, that respondents most active in their environmentalism (that is, those actually making monetary contributions to ecology organizations)—a total of 22 percent—all earned $50,000 a year and up.

Large marketers see this trend as a welcome challenge to increase sales. In most cases, they are honest in their attempts to provide environmentally friendly products. The confusion enters because we do not, as yet, have a full set of scientific and legal standards by which to judge "environmentally friendly" products.

At the beginning of 1990, only about eight states were investigating strong and uniform pollution standards, including the real meaning of *biodegradable*, an often-abused term in marketing. Numerous plastic and paper products are, indeed, biodegradable—if they are exposed to air and sunlight over a period of time. However, landfills bury waste-filled bags vertically. No air or sunlight reaches any of the waste except what reposes temporarily on top. In Arizona, an ''archaeological'' dig was made into a large landfill. Going down into a gigantic pile several years old, plastic bags and old newspapers a decade old were unearthed—still quite readable.

Plastic recycling is too young a field to provide objective scientific data on decomposition—despite the claims of Mobil's Hefty trash bags and First Brand's Glad trash bags.

Opportunities in ecological hazards

Pogo was right, no one is without blame. With recognition of the problems, however, comes a will to find solutions—and opportunities for the right types of entrepreneuring. So far, this chapter has touched on just some of the things we are up against. The list will grow as you read the chapters that follow, but the opportunities will always be apparent. The remainder of the chapter covers 10 entrepreneurial opportunities created by ecological hazards.

Rising temperatures

Rising temperatures, or the ''greenhouse effect,'' refers to a buildup of gases, usually carbon dioxide, in the atmosphere. This gas forms a canopy or top of a greenhouse, preventing the gas close to the earth from escaping like it normally does. Rising earth temperatures cause some of the ice caps to melt, raising the level of our oceans and potentially inundating low-lying coastal areas. Opportunity: to control industrial carbon dioxide emissions, especially in industrial nations and major cities with heavy automotive traffic.

Deforestation

Denuding forests, especially tropical forests, reduces the absorptive ability of land masses to absorb harmful gases. Reducing tropical forests in the Amazon Basin, for instance, can affect carbon dioxide absorption and weather conditions thousands of miles away. Opportunity: controlled forest land management; reforestation projects on a massive scale, such as those done in Israel, Indonesia, and parts of the United States; and scientifically controlled harvesting of forests.

Ozone layer penetration

Penetrating the ozone layer, a protective layer of gas in our upper atmosphere that protects the earth from the full force of the sun's rays, is increasing pollution on earth. Rising masses of dangerous gases have, in the opinion of scientists, reduced or perforated the protective ozone layer. Opportunity: worldwide pollution control devices and education as to these international dangers.

Acid rain

Acid rain is the by-product of polluting particles rising from the earth and mixing with sunlight, condensing and coming back down to earth in the form of harmful rain, snow, or hail, creating acids in rivers, lakes, and oceans and killing fish, aquatic animals, and aquatic plants. Opportunity: creating pollution control devices, especially on industrial chimneys that discharge sulfur dioxides and nitrogen oxides; purifying premium grades of industrial and automotive oils by filtering out dangerous by-products; and preventing and controlling forest fires, thought to contribute to atmospheric acid rain pollution.

Air pollution

Air pollution is dirty air caused by combusting or burning fuels. Most of this pollution is caused by power plants, auto exhausts, factory smokestacks, and wood stoves. Opportunity: developing smokestack control devices, cleaner fuels (lower sulfur content), cleaner-burning fuel burners, and improved wood stoves, such as the LaFontaine Biomass Institute wood-burning stove, especially in lesser-developed nations.

Water pollution

Dumping or seeping harmful chemicals directly into water supplies or indirectly into underground water sources pollutes the water. The vast increase in factories, greater use of dangerous chemicals in production and agriculture, and the explosive accumulations of human sewage have made this problem an immense danger during this century. Opportunity: developing mechanical control and filtering devices, processed water for home consumption, substitute drinking supplies, and improved chemicals.

Hazardous and toxic wastes

Discarding, dumping, or leaking dangerous chemicals, or even burying them in dumps near human water supply sources, makes up hazardous and toxic wastes. Opportunity: locating, transporting, and burying chemical waste products safely; and manufacturing spill-proof and erosion-proof containers.

Solid wastes

Solid wastes include trash and garbage that cannot normally be recycled economically. The average American family of four generates more than 2,500 pounds of garbage each year (65 million families would generate over 160 billion pounds of refuse annually). Opportunity: develop recyclable products and new effective ways of solid waste disposal and trash removal (which has become a major industry as well as a financial success).

Pesticides

Pesticides can kill insects and undesirable rodents and often, aid in the greater growth of crops, but unwisely used, such chemicals can cause widespread problems for wildlife as well as humans. Opportunity: developing of effective but safe pesticides; learning proper application of chemicals; and creating safe pesticides for consumer and agricultural industrial applications.

Soil erosion

Soil erosion is when land is cleared for agricultural and construction use or overusing soil so that it loses its ability to grow crops, plants, or trees, and can be blown away by winds. Opportunity: develop reforestation, horticultural, and landscaping projects that counter the erosion; soil testing; and hydroponic growing.

Obviously, there is no lack of opportunity for those entrepreneurs who have imagination, knowledge, and experience, and the fervor to reverse the ecological abuses of past generations.

2

Alternative fuels: reducing pollution

FOSSIL FUELS—COAL, OIL, AND NATURAL GAS—ARE NOT RENEWABLE. Geologists have reasonably accurate tables as to when these sources will expire. It is inconceivable that a day will come, well beyond our lifetimes, when the Alaskan pipeline will be dry; when Arab sheikhs might go back to becoming beduins; and when Mexican, Venezuelan and North Sea oil platforms will become diving boards for Sunday shore excursionists.

When this happens, the sun will continue to shine, the wind will continue to blow, ocean tides will continue to ebb and flow, geothermal heat will continue to vent its energy to the surface, and vegetable matter will continue to compost into biomass fuel. If all of these sources are available to us forever, why not start now to develop reliable surrogates for vulnerable fossil fuels?

There is one unlimited source of energy available just for the processing. No digging out of the earth; no polluting the environment; no uncertainty about the political situation in the Middle East; no fear of oil spills; and no economic variables from one month to the next. It's solar energy.

So how come we don't concentrate our technology on the power of the ever-present, free-for-the-taking sun? How come 4 million homes in Japan use solar heating? Why are more than 70 percent of Israeli homes and 90 percent of homes in Cyprus equipped with solar hot water heaters and perhaps only 1 million in the United States? The culprit is low-priced oil.

Solar energy

In the eighties, when oil was high-priced and uncertain, American ingenuity latched on to the sun in earnest. The government added impetus by allowing residential, renewable-energy tax credits. All over the country, small entrepreneurs jumped into the solar heating business and convinced nearly a million home and pool owners that they could heat their premises with solar collectors—and not worry about tomorrow's OPEC whims or the mood of the Arab sheikhs who controlled the oil spigots. Then, in 1986, the world economy settled down once more and oil prices plummeted—along with solar installations. Instead of encouraging the fledgling industry and the other 64 million home owners to give limitless solar energy a chance, the Reagan administration dropped the solar ball and solar energy went back into its shadow.

Those home owners that installed solar heating before 1986 have the system almost paid off in savings, however. It takes 6 to 10 years to pay back the initial costs (it could be amortized over a comparable period)—the variable depending on local energy costs. Today, improved installations are available that can pay back the initial installment costs in as little as three years. Solar system designers are becoming more sophisticated, too—providing choices between active systems, passive systems, utilization of photovoltaic cells (composed of inexpensive, prolific silicon), and solar-thermal generation. Technicians and scientists predict that readily available and economically viable solar energy will be here within 50 years. Why not get a head start now?

Take wristwatches and pocket calculators, for example, which have been around since about 1978. Their paper-thin photovoltaic films have found their way into increasing numbers of consumer products and their prices are dropping steadily. Best yet, they work in Duluth, Minnesota, as well as in Miami, Florida.

A home would require a much larger panel than a wristwatch, however, say an 18-square-foot panel, for instance. Operating at only 20 percent efficiency, it would still save an average of 1 kilowatt annually—enough to provide electrical power for the year, even in areas of intermittent sunlight. Obviously, Southern California, Florida, Arizona, New Mexico, and most of Texas would be much more efficient locations for PV power. Caribbean islands provide proof positive that, when sun and savings combine, solar energy has a good chance to prove itself.

Sun-fueled homes

As an example, a test installation was erected in the Dominican Republic in 1984 by Enersol, a nonprofit organization out of Massachusetts, with a grant of $20,000. The first unit was installed on a home-and-shop in a small rural town. It was a 35-watt solar module that generated enough electricity to light the premises and run the radio and television set—at a grand cost of $400. The following year, an entrepreneur opened a hardware store there and began to handle the installation of other solar-powered projects. To date, more than 600 have already been installed, and the total is climbing by 400 to 500 annually. Each unit produced about 35 watts of power at a cost of around $500 and $23 a year for biannual battery replacements. The Dominican entrepreneur is doing alright, too. (For details on this project contact Mr. Richard D. Hansen, Enersol Associates, One Summer Street, Somerville, MA 02143.)

Abundant sunlight in the American Southwest has also spawned many modest installations. The Laguna Del Mar subdivision in San Diego, California has well over 100 homes equipped with PV modules as well as a backup system tied into the San Diego Gas and Electric Company. A 6.5 megawatt plant was opened in this area in 1985, at this time, the largest in the world. Located in Carissa Plains, California, it is operated by Arco Solar and Pacific Gas & Electric Co. A far larger installation is under construction and ready to open in 1992. The Chronar Corporation and SeaWest Power Systems, San Diego, are joint-venturing this huge, 50-megawatt PV plant.

Best of all, such solar energy installations have virtually zero or very little environmental impact. Its entrepreneurial implications are vast and limited only by man's imagination and economic realities.

Chronar Corporation has also manufactured more than 800,000 walkway lights, all powered with PV cells. They have graduated from these lamps to billboard and parking lot illumination, using high-efficiency PV-powered metal halide lamps.

The U.S. Coast Guard also uses this technology in more than 20,000 navigational aids and lighthouses. More installations have been contracted for by the National Park Service, the Department of the Interior, and the Department of Agriculture. Transportation and security companies are also finding the technology useful and reliable, as well as cost-effective.

Recent experimentations by the Department of Energy at their Sandia Laboratories near Albuquerque, New Mexico have led to another interesting development. Instead of the current 20 to 21 percent efficiency for PV power, they have reached 31 percent and are predicting 35 percent efficiency in the near future.

Home builders and potential owners of custom-built homes can obtain more information on solar energy potentials from the Solar Energy Industries Association, 1730 N. Lynn St., Suite 610, Arlington, VA 22209-2009, or get a free booklet called *Tips for Energy Savers* from the Department of Energy, Conservation and Renewable Energy Referral Service, P.O. Box 8900, Silver Spring, MD 20907.

Environmentally speaking, remember that a single residence equipped with a solar, hot-water heater can eliminate nearly 1,500 pounds of atmospheric carbon dioxide emission in just one year.

From the entrepreneur's viewpoint, solar energy today is a $500 million industry and is expected to more than triple by the end of this decade. Those who get on the sunshine bandwagon are sure to enter a bull market, as well as contribute to the purification of the earth.

Sun-fueled transportation

Solar-powered transportation is also off the drawing boards. Early in 1990, a foam-balsa-and-Mylar contraption that looked a little like an airplane was *pedaled* over the floor of the San Joacin Valley in California for a little more than a mile in just eight minutes flat. At the controls of the ingenious plane was pilot Bryan Allen, a modern-day Wright brother. It was the latest in a series of experimental designs by Paul B. MacCready, the aerodynamic genius from the California Institute of Technology who now heads the AeroVironment Corp.

From an airborne vehicle, it was just a short hop to a land-based one, powered by solar energy. In 1987, AeroVironment produced the *Sunraycer* to compete in a grueling 1,950-mile, two-and-a-half day race across Australia—and won. Today, this vehicle can be seen in the Smithsonian Museum in Washington among other icons of American transportation achievements.

Electrical energy

Currently, solar power does not seem to be marketable for automobiles. Electric propulsion, however, is on the verge of reality. The sun-powered car that won the cross-Australia race did not exactly streak across that continent. Its top speed was 49 mph. The *Sunraycer*, however, became a prototype for the GM-planned *Impact*. The most advanced plans for an electrically powered vehicle.

In 1938, GM produced another electrical-powered prototype but it never got into showrooms. Then, as now, the big bugaboo is the battery—or rather, a battery of batteries. The *Impact* will need 32 batteries that weigh 870 pounds. They will need to be recharged every 120 miles and replaced every 20,000 miles—which means, on average, every two years.

According to GM chairman Roger Smith and chief engineer Alec Brooks, it will be a lucky day if the *Impact* goes into production as early as 1991. It is planned to be introduced first in California and priced in the neighborhood of $20,000. The plastic two-seater *Impact* is planned to weigh in at 2,300 pounds, enabling it to hit 60 mph in just 8 seconds and reach a top speed of 110 mph. There are two problems that enthusiasts—and environmentally concerned drivers—should expect. One, the cost of driving a clean and quiet car will be about double that of today's gas-powered vehicles. The other problem? Look out! The *Impact* will be so quiet that pedestrians won't hear it coming.

Wind energy

Wind power has been harnessed with some success. In California, a "wind farm" has been running for some years and producing commercial quantities of electricity in outlying areas. For those living in rural areas, a single-unit wind turbine can generate up to 10 kilowatts, sufficient enough to power an average household.

The entrepreneurial opportunity of wind energy, coupled with its environmental benefits, is available to anyone who takes the time to learn wind technology. It has been around at least since 1900. At the turn of the last century, about 5 million water pumps driven by windmills dotted the American landscape. The fuel was nonpolluting and it was free. Metal-bladed water pumpers, costing about $1,000 or a little more, still exist on many farms today. An annual lubrication is all the maintenance they need.

Progress in wind-powered energy producers—at a cost of 2 to 3 cents per kilowatt—lasted well into the 1930s. For 40 years, rural electrification, fueled by cheap coal and oil, reigned and displaced wind power. By the mid-1970s, new wind-energy technology and the Arab oil embargo revived wind power as a viable alternative.

The new generation of wind turbines range all the way from remote-location 1 kilowatt windchargers to 175-foot-tall towers holding 150 foot blades and producing megawatts of power but at an installation cost of about $4 million each.

In order for wind power to be effectively and economically harnessed, winds must blow at 10 mph or better, usually along seashores, on islands, or in mountain passes. California's coast, the coast of New England, the Pacific Northwest and Hawaii are where most present-day wind turbines are located. California's wind farm has as many as 500 of these tall turbines, each generating enough electricity to equal the output of a conventional coal- or oil-fueled plant.

Experimentation has found that wind farms in the 200- to 500-kilowatt range are most efficient and economical. Many of the best ones are manufactured in Denmark and Belgium. With California—Altamont, San Gorgonio, and Tehachapi as primary locations—and Hawaii in the lead of this technology, governments should be reminded that California's goal is to produce 4,000 megawatts of wind-generated electricity by 2000 A.D. at an estimated savings of 25 million barrels of oil a year!

Despite the decline in wind power installations during the Reagan years, the industry is making a strong and unheralded comeback. The 150-wind turbines in 1981 have grown to 18,000 in 1990, generating nearly two billion kilowatt hours of electric power, enough to power 300,000 homes and save 3 million barrels of oil.

Another attention-getting comparison can be made for the 500-tower California wind farm, which produces power equivalent to the output of two large nuclear plants. The wind turbines were built in half the time it takes to build a nuclear plant and at half the cost of a conventional power station.

It is estimated that by the year 2000, we can have more than 32,000 megawatts of wind-generating equipment in place. The benefits would include a drop from 7¢ to 4¢ per kilowatt hour, a reduction of 1,150,000 tons of sulfur dioxide (SO_2) and 50,700,000 tons of carbon dioxide (CO_2) emissions. The payback, obviously, is tremendous.

Geothermal energy

The very part of the earth that is home of the largest dry-steam development in the world is also the unpredictable host to the largest population concentration in the United States—California.

Operating since 1960, the Geysers steam field, 90 miles north of San Francisco, is about half of the world's discovered natural, dry-steam wells. At present, only a tiny fraction of this underground energy is recovered.

The Geyser fields in Northern California are only a part of the tapable natural energy source. Southern California's Imperial Valley has sufficient underground hot-water resources to meet the electricity needs of the entire southwest United States.

With the application of modern technologies, the California Energy Company has been able to increase geothermal capacity from 27 megawatts to more than 200 megawatts in the span of only two years between 1987 and 1989. The emissions from CO_2 and SO_2 are virtually zero. This environmentally acceptable energy production is being accomplished more cheaply than comparable energy derived from fossil fuels or nuclear plants.

There are other, less efficient underground reserves along the Gulf and Eastern seaboard areas. The Electric Power Research Institute claims that with current technology, North American geothermal power capacity could reach 20,000 megawatts by the end of this decade. One large system in Italy has been generating commercially usable quantities of electricity since 1904—but as with other sources of alternate energy, it will take concerted national policy and effort to relinquish total dependency on fossil fuels and nuclear energy.

The opportunities for "green entrepreneurship" are truly hot, even if they are, at this point in time, still mostly underground.

Hydropower energy

It might come as a surprise, but current hydro capacity in the United States is close to 90,000 megawatts—13 percent of total electricity generation. To this can be added another 60,000 megawatts, which is generated in Canada and is available to us. One large plant in Southeastern Canada produces 10,000 megawatts, much of which powers New England homes and factories.

Not counting the substantial capital costs, production of electricity through hydropower costs the end-user between 3 and 6 cents per kilowatt hour. This compares most favorably with other resources. All told, North American plants produce 30 percent of the world's output and the potential for hydropower's expansion is virtually limitless.

The World Bank, which finances many hydropower developments in other countries, such as China, India, and Brazil, estimates that, by 1995, at least 223 gigawatts (GW) of electricity will be produced by the new generation of super-plants. As an example, Venezuela already completed what is, in 1990, the world's largest hydropower dam in the world. It has a generating capacity of 10 gigawatts, while the output of a modern nuclear power plant produces only one gigawatt. Also on the drawing board or in the planning stage is a 12-gigawatt plant in Brazil and a 13-gigawatt plant on the Yangtze River in China.

Large dams of this type are not ecologically sound, as was proven by the Aswan Dam in Egypt, and the Three Gorges dam in China. The experiences in the United States also set a sound example. The most effective and ecologically preferable projects fall into the range of 100 kilowatts to 10 megawatts (small) and 10 megawatts to 100 megawatts (medium) size.

In the United States, we have, as yet, much undeveloped hydropower potential. The Ohio River, for instance, presently produces only 180 megawatts of electricity, but its potential is more than 15 times that capacity. By comparison, the Rhone, a major river in France, produces at least 3,000 megawatts of power through its 21 dams.

Harnessing the tides

France and Norway also have used the tidal power of water along their coasts for a score of years. The technology has been made available to Indonesia, Portugal, and other countries with access to tides. A Norwegian tidal power installation was constructed for less than $2,000 per kilowatt, generating electricity at a cost of 5 cents per kilowatt hour.

Because of low operating costs, tidal hydropower plants have tremendous long-term potential for the generation of electricity with far lower ecological impact than large dams.

Biomass

Biomass is a general term for any natural product that grows by photosynthesis. It is a renewable energy source derived from waste wood, crop wastes (sugarcane, rice husks, etc.), animal wastes, and organic garbage. It can come in the form of liquids, gases, or solids. Only 15 percent of available biomass is used for energy worldwide, 8 percent by industry in the United States. The only state in America that uses any appreciable amount of biomass in the production of electricity is Maine, which utilizes 23 percent, or 640 megawatts, of its capacity.

Biomass does not necessarily mean cutting down trees to burn them for energy production. About 25 percent of our landfills, it is estimated, is composed of wood wastes. Half of the world's population, however, is dependent on wood for cooking and heating. In India and sub-Saharan countries of Africa, this practice has already led to serious deforestation and soil depletion. As a contrast, large wood by-product-producing countries like Scandinavia, Russia, and Canada use these biomass wastes in cogeneration systems to produce both electricity and heat. In the United States, four wood-burning power plants exist, producing more than 200 megawatts of electricity combined.

In Hawaii, especially on the agricultural island of Kauai, 60 percent of electricity is generated by biomass plants—utilizing waste products from sugarcane, rice, coffee, cotton, and coconut plants. In the Imperial Valley of Southern California, about 100 miles east of San Diego, farmers are selling manure at $1 a ton instead of paying a refuse company to haul it away. Southern California Edison buys the stuff and now generates enough electricity to service 20,000 homes—saving 300,000 barrels of oil in the process. In addition, it should be noted that resultant electricity is sold profitably for 7 cents per kilowatt hour, and the residual ash from the incineration process—150 tons a day—is recyclable as fertilizer and as an ingredient in paving mixtures.

Biofuels are also usable for the production of methane, ethanol, or methanol. In Monroe, Washington and central West Virginia, biogas plants produce as much as 100,000 gallons of fuel in a cost-effective recycling system. Wood alcohol, a premium fuel, can also be derived from biomass conversion through fermentation and distillation.

Of all biomass fuels, plant-derived ethanol is more efficient and less polluting than methanol. The 1988, U.S. Alternative Fuels Act sanctioned credits to automobile manufacturers who produce vehicles utilizing fossil-fuel saving, less polluting engines. These credits become effective in 1993 and result in the manufacture of 100,000 alternative-fuel cars. The 5,000-foot-high city of Denver, Colorado already has required cars to switch to gasohol, especially during the winter when pollution is at its worst. Tests have shown that carbon monoxide emissions have dropped as much as 20 percent.

The bad and the good

A decade ago, the federal government spent a billion dollars on alternate energy research and production—almost as much as on nuclear fission. Then oil prices started their downslide on the world market and we again took the easy way out. Today, hardly $100 million of federal money goes into alternate energy research and development. While solar power is perhaps the greatest and most reliable source of alternate energy, other alternates are available. It is unlikely that, based on our present technology, solar power alone will suffice to make up for a major decline or exhaustion of fossil fuel energy. But, between biomass, hydropower, geothermal power, wind power, and solar power generation, the United States can yet beat foreign oil suppliers at their game—and help keep our air breathable, not to mention make a lot of money in the process.

3

Up in smoke: opportunities for cleaner air

WHAT WE DO DOWN ON EARTH CAN WREAK HAVOC IN THE SKY ABOVE. We have the potential for damaging the planet's ozone layer—the shield that protects us from the effects of harmful, ultraviolet radiation. We have the potential to put enough carbon dioxide into the air that the earth's atmosphere warms and causes devastating changes in our climate. We also have the potential to stop what we are doing—to find alternatives for the harmful chemicals we are using and find golden entrepreneurial opportunities while we're doing it.

Chlorofluorocarbons

We have all heard about chloro—what's-its name. In fact, for a few decades, we have taken it for granted. Chlorofluorocarbons (CFCs) are man-made gases and solids that we use in propelling hair spray and many other sprays and cool our refrigerators and automobile air conditioners. It is also the stuff that could put this earth, as we know it, out of business. Including us. So let's take a look at what CFC and its siblings really are, what we can do about them, what alternatives we have, and how we can help save the—no, make that *our*—environment while turning a good buck.

CFCs are relatively nontoxic, nonflammable, stable compounds. They come in eight different forms and go by the family names of either CFC or Halon. Their virtually unpronounceable full chemical names are as follows:

CFC-11	Trichlorofluoromethane
CFC-12	Dichlorodifluoromethane
CFC-113	Trichlorotrifluoroethane
CFC-114	Dichlorotetrafluoroethane
CFC-115	(Mono)chloropentafluoroethane
Halon-1211	Bromochlorodifluoroethane
Halon-1301	Bromotrifluoroethane
Halon-2402	Dibromotetra fluoroethane

If you buy, or want to produce, any product that contains any of these chemicals, put it back on the shelf and run. These products are contained in aerosols; foam plastics; aerosol dust removers; some fire extinguishers; building insulation materials; cleaning sprays; boat horns; VCRs and other electronic equipment; and much plastic, foam food packaging, and more.

What makes CFC compounds so dangerous?
In gaseous form, or when burned, CFCs are carried upward into our atmosphere. About 6 to 30 miles straight up, they stop and form an invisible shield. But there is already a shield up there called the ozone layer. This is the good layer that helps screen out the ultraviolet rays of the powerful sun. Overexposure can be harmful, however, these UV rays can cause skin cancer, sunburn, and topographic dehydration. The Germans have a saying, "Zu viel ist ungesund"—meaning, "Too much is unhealthy." And that certainly applies to getting too much UV exposure.

UV exposure can be dangerous to your health, although a little bit is good for you. The proper amount of UV rays are controlled by your common sense and by the protective ozone layer, 6 to 30 miles up in the sky. Now, imagine that some of that ozone layer is diluted or even removed. An extreme scenario would say that human life as we know it would not be possible.

Scientists have long suspected that CFC gases damage the ozone layer that shields us. They even undertook a number of expeditions and found that a puzzling "hole" had developed over the Arctic region. Experimentation proved that CFC gases that rise into the atmosphere are broken down chemically by exposure to large doses of ultraviolet radiation, a process that frees the chemical chlorine contained in all CFCs. Once freed, a chlorine atom destroys about 100,000 molecules of ozone, and then begins to drift downward again where, perhaps several years later, it lands on the earth's surface.

Thus, CFCs affect not only our own well-being, but depress human and animal immune systems, reduce crop yields, and reduce fish populations. CFC is quite democratic. It affects every human being and every human habitat. Being very stable; it is estimated that CFC atoms last as long as 150 years. Therefore, we better not give them that chance, even if *we* aren't going to be around that long.

Times—they are a-changing

A great number of restrictions against CFC use are already in place. Plastic foam packaging, for instance, has already been banned in Portland, Oregon; Tempe, Arizona; Newark, New Jersey; the entire county of Suffolk, New York; the entire state of Florida; and other cities in California, including Berkeley, Los Angeles, San Francisco, and Palo Alto. The state of Vermont will have a CFC ban used in car air conditioners.

Irvine, California, to take an example, a city of about 60,000, enacted a law in July 1990 that prohibited the use of virtually all CFC products in any industrial process except drugs and medicines. Even garages and service stations have been mandated to recapture and recycle used Freon and other CFC compounds. There will be no more foam cups, foam packaging, or foam building materials. At least the sky over Irvine will be relatively ozone-whole.

With more restrictions on the way, clearly it behooves businesspeople to get out of such sensitive and soon-to-be illegal merchandise.

CFC substitutes

Chemists believe that man-made CFC compounds can be eliminated by developing safe substitutes. Take the electronics industry, for example. They replaced a harmful, CFC-dependent spray cleaner that was used to clean electronic components with a new solvent made from water and detergents, or even from orange peels, that does the job just as well. Coolants in new refrigerators and air conditioners also use safer compounds and even with new safety precautions, are safer than old-fashioned ammonia.

An even greater discovery for environmentalists and entrepreneurs is that these alternate compounds have proven even more efficient. The Environmental Protection Agency announced that the greater energy-efficiency of CFC alternatives could save us $5 billion in energy during this decade alone. Better yet, this energy savings will accelerate in the next century to over $100 billion—not even including the substantial health savings.

Entrepreneurial opportunities

There are infinite opportunities for entrepreneurs who catch the energy-saving, anti-CFC movement. For example, converting millions of home appliances—air conditioners, refrigerators, freezers, water heaters, washers and dryers, and dishwashers—to alternate compounds could result in billions of dollars in sales as well as an estimated $50 billion in energy savings by 2010. Other businesses that could be further developed on the heels of this anti-CFC trend might include:

- Recycling CFC products from garages, service stations, repair shops, major appliance repair shops, and dealers.
- Safely salvaging old CFC refrigerators and air conditioners.
- Promoting the use of fiberglass, cellulose, and other safe materials as building insulation.
- Promoting the use of fire extinguishers that use dry chemicals rather than CFC-Halon compounds.
- Finding products that substitute CFC-activated aerosols and using and promoting their sale.

The greenhouse effect

More than 10,000 years ago, a gradual warming of the earth melted several hundred feet of ice where Chicago stands today. Lake Michigan is one of the results. Perhaps dinosaurs disappeared some millions of years before because of just such a global climatic change. If the globe continues warming as scientists are predicting, will there be any coastal cities in the United States left high and dry? Indubitably not.

The "greenhouse effect" is just that: a gradual warming of our atmosphere, causing a slow melting of the polar ice caps and a raising of the water level all over the world by as much as 50 meters. "Washington Under Water" makes a plausible sci-fi thriller but should there be a scintilla of truth in the scientists' conjecture, shouldn't we look into it?

What science has to tell us

Scientists from NASA, the University of Illinois, Goddard Institute for Space Studies, Stoney Brook State University of New York, and the Woods Hole Oceanographic Institute are just some of the minds wrestling with this problem. Computer models, taking into account the 25 percent increase in atmospheric carbon dioxide since the advent of the industrial age, predict a 3 to 9 degree Fahrenheit surface warming over the globe during the next century.

Some scientists feel that the warming in the next century will be the equivalent to the warming that took place over a 10,000-year period following the last ice age. Others refuse to admit to such a possibility because we do not, as yet, have infallible scientific evidence.

We are still waiting to hear, incontrovertibly, whether such dire predictions as a flood-producing earth-warming will take place during the next generation or the next century. While a hundred years are a blink of an eye in geophysical terms, the human mind is ill-equipped to think 100 years ahead in pragmatic terms.

A hundred years can bring remarkable technological and scientific changes. Think what the world was like in 1890, for instance. Horses pulled carriages and trams while 30-mile-per-hour trains chugged westward. Motorized transportation through the air was a Wellsian dream.

Sixty-three million people inhabited the United States and the center of the country was a little west of the middle of Indiana. In the east, 400 people died in a March blizzard and 3,200 more in the Johnstown, Pennsylvania flood. Was this a signal of weather changes ahead?

Politicians get into the act

The greenhouse effect also carries repercussions in the political arena. It is dramatic and exciting and just the stuff that colors political speeches. A scientific writer and philosopher, Alston Chase, had this to say about the disturbing pattern in the way public policy is formed on environmental issues:

''It begins with environmental groups making claims, often premature and sensational, that represent only one part of scientific opinion. The media pick them up and a bandwagon develops which sweeps many grant-seeking scientists along with it. That same bandwagon can also sweep the nation into costly commitments made, not in the light of reason, but in the heat of politics.

''It is the nature of our minds, and the press that is reflective of *vox populi*, to make headlines out of a story that shouts ''Global Warming Will Put U.S. Coasts Under Water'' but that the story to the contrary, ''Global Doomsday Cancelled'' will never make it to the front page.''

Controlling CFCs and CO^2

So where does the entrepreneur come in with all this political and scientific waffling going on? Combatting the ''greenhouse effect'' is like carrying car insurance or putting on a seat belt while driving. It is carrying homeowners insurance against fire, earthquake, and floods. Of course, you don't expect your house to catch on fire, your car to be totalled in a traffic accident, or you to die before your time is up at a ripe old age. But just in case you're wrong, you take out an insurance policy and you pay premiums for most of your life.

For the same reason, we ought to pay attention to the pernicious greenhouse effect. But can we do something about saving the sky? Can entrepreneurs combine their skills and ambitions with the environment's plight—and still profit from the earth? Can it be done in a Western democracy? Of course. Germany has started a national environmental plan that has already shown spectacular results. According to economists, its total energy use per gross national product unit is just about half of the energy used by U.S. industry. It plans to cut carbon dioxide emissions by 25 percent by the year 2005. These efforts translate into greater savings, greater profits, and greater competitiveness in the increasingly global marketplace.

If we want to stay healthy, stay profitable, and stay competitive, CFC and CO^2 control is an initial and achievable step. The following are just five areas in which industry and business can work to save the earth while reducing the greenhouse effect—while effecting the greening of their own houses.

- Removing most of the 6 billion tons of carbon dioxide that are spewed into the atmosphere each year.
- Eliminating the use of chlorofluorocarbons and going back to air-pressure and plunger-type dispensers.
- Developing alternate fuels, primarily for motor vehicles and home heating/cooling that use renewable energy like sun, wind, and water.
- Increasing the efficiency of appliances, illumination, transportation, and heating and cooling to reduce the wasteful and often hazardous use of energy fuels.
- Changing chemical processes and production to eliminate or drastically reduce the 2.4 billion pounds of dangerous emissions that are released into our air each year.

There is no doubt that there are a number of long-term, big-scale entrepreneurial opportunities available today. The key is capitalizing on them. Although environmental opportunities existed for the business-person long before the 1990s, today's widespread interest in the earth will begin to propel environmental businesses to new heights.

4

Environmental alchemy: recycling waste into wealth

GARBALOGY HAS BECOME A HUGE AND OFTEN CONTROVERSIAL SCIENCE and enterprise. The statistics are staggering. The average American, by the time he or she reaches a venerable 75 years of age, has produced 52 tons (104,000 pounds) of garbage, costing the municipality in which John or Jane Doe lives, $2,600 to dispose of it. It costs at least a million dollars a year to dump the garbage generated by 100,000 people in the United States.

Burning is no solution because 30 percent of all garbage incinerated ends up as toxic ashes, the disposal of which costs vastly more than getting rid of the garbage in the first place. It costs $50 a ton to bury our garbage in a landfill, and as much as $75 a ton to incinerate it. Recycling's bill, on the other hand: $30 a ton.

Manufacturers must take responsibility

When we examine the 3.6 pounds of waste that each of us generates each day, it is no surprise that much of this trash comes from packages of food, tools, appliances, garments, cosmetics, and all the trappings of our advanced civilization.

Consumer packaging is responsible for a major portion of our waste products and its removal cost has not been lost on government circles.

Landfills are reaching a critical stage. In another year or so, Connecticut's dumping grounds will be filled. New York state is expected to reach 100 percent saturation by 1995. New Jersey, running out of landfills, is "exporting" its surplus garbage. The American trash can is just about full and the smell is not pleasant.

The Environmental Protection Agency in Washington has taken the lead in identifying the many-sided problem. "Manufacturers have no direct incentive to design products for effective waste management because they are not usually responsible for the costs of waste management," stated the Environmental Protection Agency's report containing goals for 1992. The good news is that the Environmental Protection Agency has added some teeth to the 1992 goals by making manufacturers more responsible and responsive to the high cost of ecological cleanup. The Agency's new regulations will include:

- Reduced packaging that will diminish waste by at least 25 percent by 1992.
- Increase overall waste recycling efforts to 25 percent.
- Incinerate 20 percent of the remaining waste in waste-to-energy plants (today less than 10 percent is processed through such incinerators).
- Utilize landfills only for non-recyclable, unincinerable materials, and for the ashes that are the presently necessary fallout from incineration plants.

A few municipalities, like Hamden, Connecticut, are beginning to enact ordinances with real teeth in them that should give the originators of waste products, the manufacturers and packagers, some concrete incentive. That push is provided by $100 daily fines and short-term jail sentences for using banned polyvinyl chloride and polystyrene, nonbiodegradable wrappings.

Criteria of civilization? Rubbish!

We are all familiar with periodic reports from the Holy Land, Hopi Indian reservations, Mexico City's subterranean diggings, and China's underground burial grounds. Archaeologists are down in the dumps, digging up the garbage of past civilizations in order to study it and recreate past history from their findings.

In 1984, a garbage project began outside of Phoenix, Arizona in Tempe. Similar digs have been conducted, all quite scientifically, in Tampa, Florida and Staten Island, New York, as well as in San Francisco. Newspapers made up about 16 percent of the "historic" samples, some quite readable. One dump revealed a 1948 plastic bottle and a 40-year-old broom with most of its bristles intact. Biodegradability was evidently not as certain as originally thought. Perhaps the word needs to be restructured into *photodegradability*—but this is but a matter of semantics. Our colossal mountains of waste products could hardly be spread out so that sunlight and oxidation would decompose it. Down in the dumps, in the dank dark bowels of the landfills, garbage is preserved, not decomposed.

What we will no doubt have to do in centuries to come is what most other civilizations have done—build on top of our refuse and give the archaeologists of 3000 A.D. some real thrills—and puzzles.

Incineration, the not-so-good alternative

Despite the high cost of incineration and despite the toxic fallout, burning our solid wastes continues at an accelerated pace. The reality of the situation is that we are running short on landfills and must have short-term solutions or we will drown in our own affluent effluent. Two entrepreneurial companies have provided partial solutions, each in somewhat different directions.

Ogden Projects of New Jersey has been operating incineration plants since 1986, burning large volumes at ultra-high temperatures and, in the process, producing electricity and steam which is sold to local utilities. Ogden has contracts with numerous municipalities which makes the company relatively risk-proof. Its nearly $25 million profit for 1989 is virtually guaranteed for each 20-year contract. Annual future growth is expected to be about 30 percent.

Ogden's elder competitor, Wheelabrator Technologies of Massachusetts, started in the incinerator business in 1975. Along the line, they acquired the refuse-hauling giant Waste Management, Inc. This company also operates recycling plants (mostly paper mills) as well as cogeneration plants and incinerators. The latter focus mostly on commercial contracts—merchant plants—rather than municipal ones. While its profit is higher, their returns are less stable because the commercial market tends to fluctuate more.

Both companies' state-of-the-art incinerator plants will probably have little trouble adapting themselves to the Environmental Protection Agency's stricter emission controls. By 1994, so the Environmental Protection Agency edict goes, no more batteries can be burned, 25 percent of the intake must be recycled, and 90 percent of the emissions must be controlled.

It will cost even more money to burn our trash in the years to come, but so far, commercial and municipal customers are willing to pay the fare in order to protect the environment. This is welcome news for environmentalists. Still, getting into the "garbage into gold" business takes an enormous amount of money. Recycling, on the other hand, can be a business for the small entrepreneur and nonprofit organization.

Recycling: turning trash into cash

Recycling waste products sounds like the dream solution of our crowded earth's surplus products problems. Depending on which article you read, recycling is either the heaven-sent solution to the problem of finding scarce landfill space—going at $100 a ton in some overcrowded municipalities—or the source of a whole new set of supply-and-demand problems. Still, there is little choice. Recycling is not the end solution to our refuse problems, but *one* of the solutions that must be tackled simultaneously with several others.

There are three primary groups that are concerned with recycling: (1) large corporations who are in the business of waste disposal or others whose waste by-products are making money for them; (2) municipalities or other governmental divisions; and (3) the many little entrepreneurs who, by design or accident, have discovered a profit niche in the recycling business.

Case histories: two giants of waste disposal

Among the industry giants of waste disposal are Browning-Ferris Industries Inc. of Houston, Texas and Waste Management Inc. of Oak Brook, Illinois. Between them, these two companies control about 20 percent of the trash disposal business. The two top executives of Browning-Ferris are William D. Ruckelshaus and Philip S. Angell, both former officials with the Environmental Protection Agency. As a result of these leaders, this industry has become more scientific, sophisticated, and environmentally attuned. It also has made it more difficult for the small old, bulldoze-and-burn solid waste disposal operator to remain in this field.

Opening up a new landfill these days requires an investment of $2 to $4 million—more than four times what it cost in 1980. Land, of course, has skyrocketed in price, but the growth of cities has pushed available dumps further and further out into the country—and since nobody wants to have a landfill near them, whatever land is available is being brokered at premium prices. Adding further to the expense, new Environmental Protection Agency regulations demand air and water monitoring, synthetic liners, and in the future, when the landfill is finished, the operator is mandated to restore the landscape to its pristine condition or better.

It seems that one way for a small entrepreneur to make a lot of money is to develop his disposal company and landfill—if he is fortunate to find a location—and then sell it to either Browning-Ferris or Waste Management. In 1989 alone, Browning-Ferris bought 131 smaller businesses for an aggregate of $416 million, or an average of $3,175 million each. Not far behind, Waste Management acquired 97 disposal businesses in 1989 for a total of about $300 million, or an average similar to Browning-Ferris's purchases.

Waste Management's recycling operation is called Recycle America, and a Canadian branch is, of course, called Recycle Canada. Its customers include 1,350 municipalities, as well as the U.S. Capitol and the house and the senate office buildings. It is part of a $5.5 billion business.

Browning-Ferris does more than $2.5 billion worth of business, has contracts with 659 communities, and runs an operation called Recycle NOW, which handles both residential and commercial accounts. Both major companies also do curbside recycling in suburban counties.

Turning waste products into wampum

On the other end of the recycling business are other big companies like Giant Foods, a 150-store chain out of Landover, Maryland and Mobil Oil Co. Giant has been collecting newspapers since 1978. According to Barry Scher, Vice President, Public Relations Director for Giant, the company has collected more than 8,000 tons of old newspaper and sold them for $168,000. All proceeds have been used to support local charities.

In 1990, the company picked about 30 stores to test an additional recycling program—plastic pop bottles and plastic shopping bags. The bottles were fine but the bags gave them a headache. Because returned plastic bags must be clean and dry and not mixed with paper (such as cash register tapes left in the bags) before being recycled, too many deviations created insurmountable problems. In addition, people left old junk and even furniture around the collection bins, heaping additional problems upon the recycling efforts. The collections were stopped.

Now, new plans have been worked out and the recycling efforts will go on. It will, in fact, be expanded to include other supermarket chains, like many Kroger, Safeway, and A&P stores—all thanks to Mobil, which worked out the gritty problems, both human and technical.

Another large corporation, Marriott Corporation, has found that recycling office papers is a boon to saving space, a small money-maker, and even a contribution to the environment. Starting in March 1990, the company's waste paper accumulation amounted to 18 tons per month within less than half a year. Marriott arranged with U.S. Recycling of Hagerstown, Maryland to pick up its paper trash, which it sells in turn to the Fort Howard Paper Company—the very same company from whom Marriott buys all of its paper goods for its restaurants and hotels. For Marriott Corp., recycling has turned out to be a full-circle operation.

Two other entities are helping in the recycling process, and firms like Giant Foods are learning to work with them. One is municipal waste disposal agencies. The other, private refuse and recycling companies.

Municipalities as players

Los Angeles, the 3 million population metropolis, and Great Barrington, Massachusetts, population 7,000, couldn't be more of a contrast. And yet, both cities launched recycling efforts that were highly respected, but, of course, totally different. Montgomery County, Maryland is yet another example of a successful recycling program.

Case history: California

Starting in September 1990, specially designed waste disposal trucks began picking up hundreds of tons of recyclable trash from L.A.'s 720,000 households. The estimated 4,000 tons of daily household trash has city officials concerned. Half will probably wind up in dwindling landfills. The other half, according to state-mandated law, is supposed to be recycled. However, how does the city dispose of such huge quantities? The problem will grow greater when other California communities must also recycle 50 percent of their municipal waste by the year 2000.

One of the problems for cities is the cost of transportation, which has climbed as high as $18 a ton. At the same time, the market for recyclable glass and newsprint has become glutted. Processors are looking to export some of the products abroad. Stated Lee Wiegandt, president of the California Glass Recycling Corporation, "If we don't start paying attention to the end-use market now, we'll pay the piper later."

Still, some cities, like Santa Monica, California turn a neat profit on their recycling efforts. Big cities like Los Angeles must look to added taxation to raise an estimated $33 million a year shortfall in revenue. Similar problems have been encountered in large, East Coast cities. Perhaps the solution is to get more small entrepreneurs involved, as well as breaking up large cities into smaller districts. Moderate and small towns appear to fare well with the recycling process.

Case history: Massachusetts

Great Barrington, which nestles in the Berkshire Mountains of southwest Massachusetts, turned its garbage problem into a hugely successful community effort. Everybody got into the act to "train" its 7,000 citizens how to sort waste products—the Rotary Club, elected selectmen, senior center residents, church leaders, and school children, all participated. This grassroots campaign produced remarkable results.

The local newspaper, *The Berkshire Courier*, exulted in an October 12, 1989 headline: Will Recycling Work? Yup. Will It Be Easy? Nope. Great Barrington's school children and their teachers also decided that they did not want to have grown-ups messing up their planet any longer. Eighth graders got together with lower-graders and showed them how to collect and sort recyclable waste products. Then they taught their moms and dads, created colorful posters, and even decorated the town dump with recycled boughs, wreaths, and Christmas cards. The parade of local citizens to the town dump, laden with recyclable trash, looked like the Piper of Hamelin's parade.

Not to be outdone, Great Barrington's three waste collection companies put on their own acts. This included educational campaigns among householders, door-to-door talks, and a waste collection newsletter. There are no more recycling cynics in Great Barrington!

Case history: Maryland

Montgomery County, Maryland has a mandatory residential recycling program. The Recycling Coordinator works out of the Department of Environmental Protection, located in the Executive Office Building in Rockville, Maryland. Montgomery County's population in 1988 was listed as 628,000 and the number of households 235,300.

The program's goal is to recycle 27 percent of the collected refuse by 1992 and 30 percent by 1997. Already in existence is a 10-year-old curbside collection program for newspapers and aluminum cans that covers about one-third of existing homes (80,000). Drop-off sites exist at two county facilities—one taking just about everything; the other, only newspapers. Both are open for most of seven days each week.

In addition, the county has made arrangements with six, large private recyclers throughout the county. Two of them, a church and a Boy Scouts facility, take only newspapers. In addition, two supermarket chains, Safeway and Giant Foods, accept aluminum cans and/or newspapers and a number of gas stations will take in waste oil. A county telephone number is publicized where a list of such gas stations can be found.

The county sells collected refuse, such as compost, for $7.50 a cubic yard. It also supervises a county school collection program to recycle polystyrene, started in 1989. All county offices participate in a regular paper recycling program. A waste-to-energy facility is under construction and is expected to begin operation in 1992. In terms of percentages, the following breakdown of waste materials is typical: 75 percent nonfood products, 16 percent yard wastes, 7.3 percent food products, and 1.1 percent miscellaneous inorganic waste.

One facet of Montgomery County's successful recycling program is its coordinated management and promotion program. The following five brochures are published and disseminated through schools, service clubs, and libraries:

Montgomery County's Recycling Plan
Montgomery County Recycling Centers
Montgomery County Clean-Up Day
Aluminum Can Recycling Program
Help Prevent Water Pollution

Municipalities interested in learning more can contact the Department of Environmental Protection, Executive Office Building, 101 Monroe Street (6th floor), Rockville, MD 20850 or telephone (301) 217-2380. *How to Launch a Community Program* is also a useful publication issued by Conservation and Renewable Energy Inquiry and Referral Service (CAREIRS), P.O. Box 8900, Silver Spring, MD 20907. The brochure is free.

To date, at least 1,000 communities across the country have instituted curbside pickup programs. Obviously, many communities are receptive to recycling, affording entrepreneurs the opportunity to start profitable recycling enterprises in their own neighborhoods.

Establishing a workable environmental protection program or a recycling program in a community is frequently a valued and well-paid position. Figure 4-1 is an organization chart that can be used, by a professional or a volunteer coordinator to set up such a program.

Administrative office location ______________________

Population of project area ______________________

Number of households ______________________

How many in individual homes ______________________

How many in multiple dwellings ______________________

Coordinator of program ______________________

Available public drop-off sites and hours
(list different sites for each product category)

____

____

Recyclable Material Pickups

Materials	*Schedule (day, wk, other)*	*No. of Homes*
Newspapers	______________	________
Aluminum	______________	________
Leaves	______________	________
Glass	______________	________
Waste oil	______________	________
Other papers	______________	________
Yard waste	______________	________

Private recycling companies and products

____

____

Private locations for recyclable products/products

____

Organizations that cooperate

____

Fig. 4-1.

Small entrepreneurs cashing in

It is said about attic treasures that one man's garbage is another man's treasure. In the case of our enormous waste collections, evident "treasures" are the opportunities that make for countless new business start-ups and expansions.

Recycling lends itself to small entrepreneurs. New England is a particularly green spawning ground for small ecological ventures. There is Steve Faust of Montpelier, Vermont, an environmental consultant; Becky Secrest of Moore Recycling Associates, Hancock, New Hampshire; and Mike Mindell of Earthworm, Inc. who has been hauling wastepaper and consulting in and around Boston, Massachusetts since Earth Day I in 1970. Mindell is quoted assuringly, "We're hauling eight to ten tons of wastepaper every day—the equivalent of about 150 living trees." When newspaper collections are flat, white paper collected from office surplus and printers brings $60 to $100 per ton.

Ecomatrix of Brookline, Massachusetts distributes "environmentally sensitive products" in response to consumer demand. In addition to the company's commercial distribution, its executives conduct educational presentation among its customers that are designed to increase environmental awareness.

In other parts of the country there is, for example, Advanced Recycling Systems, Waterloo, Iowa. A television program on waste recycling stimulated founder Mike Nordstrom, soon joined by friends Mark Baker and Robert Armstrong, to create a series of garbage containers that contained tight-lidded bins for a variety of recyclable refuse. They make residential and commercial units and have expanded into truck-mounted compacting devices, called *Crush-R*, that have found a ready market with commercial solid waste disposal companies who have contracts with municipalities that have made recycling mandatory.

A mail order company that specializes in recycled paper products, Earth Care Paper Co., Madison, Wisconsin, was started in 1983 by a couple involved in environmental education. In the last few years, the company's sales have tripled annually. Advertising in such mainstream media as *The New York Times* has spread the environmental gospel.

In Philadelphia, Newman & Co. are reprocessing 250 tons of wastepaper a day collected from IRS offices, police departments, FBI offices, post offices, and records centers. These are all confidential papers and must be shredded. Papers that are collected from office buildings have to go through special sorting processes that include screening, centrifugal force separation, gravitational sorting, and other methods.

Turning recycling problems into opportunities

Many products can be recycled successfully. Among them are glass, steel, plastic, and motor oil.

Glass

Glass lends itself profitably to recycling. In 1989, more than 5 billion glass bottles and containers were collected, crushed, pelletized, melted, and reformed into new containers. New chemical and engineering technologies have made this process very cost-effective.

Steel

Steel is also very recyclable. Several years ago, steel processors developed techniques to recover and recycle over 56 million tons of steel and iron. AMG Resources of Pennsylvania and The Rhode Island Solid Waste Management Corp. have joined to magnetically extract metal cans from waste piles, cleanse them of all contaminants, and detoxify them before forming into pellets for reprocessing.

Plastic

Plastic, too, lends itself to recycling more and more, as new processes are developed that make re-use feasible and cost-effective. In 1990, more than 20 percent of all plastic soft drink bottles were made from recycled plastics. The much-maligned foam plastics that have proven so resistant to decomposition, will now have a second life in such nonpersonal applications as fiberfills and cushioning packing materials.

One of the major re-users of old plastics is the Hammer Plastic, Recycling Corporation of Iowa Falls, Iowa. Among such impersonal products manufactured from cleansed and reprocessed plastic pellets are oil and detergent containers, pipe racks, boat piers, chairs and benches for outdoor use, wheel chocks, and parking lot speed bumps.

Motor oil

Draining old oil from millions of cars, half the amount done at home in garages and driveways, offers opportunities for monstrous pollution. Often, old oil is dumped into storm drains and sewers where it has a chance to seep into the ground and pollute underground water sources. Collecting and recycling old oil is a profitable business, especially as oil prices continue to rise and make this operation doubly worthwhile.

Other materials

Not all products can be recycled of course. Some cannot be for physical or chemical reasons; others for cost-effective reasons. Consulting with local waste disposal officials can be a good beginning, however. There are at least a dozen chemical products that can be recycled. Collecting some of the following waste products can get any entrepreneur into the recycling business while becoming an environmentalist-par-excellence at the same time:

- Automatic transmission fluid
- Batteries and battery acid
- Diesel fuel

- Fuel oil
- Gasoline
- Kerosene
- Paintbrush cleaner made with solvent
- Paint thinner
- Lye-based paint stripper
- Turpentine
- Dry cleaning solvents
- Gun cleaning solvents

The best places to go into business

As a guide to where the best opportunities for a recycling business are located, take into consideration that, thus far, 14 states and the District of Columbia have passed and implemented bottle bills that collect as much as 10 percent of the waste that is represented by glass and aluminum containers. These 14 states are: California, Connecticut, Delaware, Florida, Maine, Massachusetts, Michigan, New York, Iowa, Maryland, Pennsylvania, Oregon, Rhode Island, and Vermont. In addition, some restrictive ordinances have been put into action in Berkeley, California, Minneapolis/St. Paul, Minnesota, and Suffolk County, New York.

One valued source of information for entrepreneurs looking for opportunities and locations, in community recycling efforts, is the *Directory of Local Governments' Recycling Practices*. It can be obtained for $10 from The Council of Governments, 777 North Capitol Street, Washington, DC 20002. You might also check your local library system.

Where do we stand today?

Only about 10 percent of all recyclable waste is actually reprocessed. Japan, which has less space to waste, nor money to burn, is reputed to recycle a full 50 percent of its recyclable waste products. And there's another beneficial fallout to recycling our garbage output. According to the Environmental Protection Agency, if all municipalities could agree to recycle recyclable glass, paper, plastics, and metals, thus saving millions of cubic feet of costly landfill space, they would save enough money to fill more than 15 million automobiles with gas for an entire year (over 10 billion gallons). At only a 10 percent profit, recycling entrepreneurs could cash in on a $1 billion business!

5

Environment-friendly investing: the green portfolio

A FUNNY THING HAPPENED ON THE WAY TO THE INVESTMENT PORTFOLIO. Bankers and other financial people suddenly saw a strange new light at the end of the profit tunnel. They perceived not stock issues, but environmental issues that made sense. Not just common sense, but uncommon dollars and cents.

Increased awareness in environmental problems and solutions have created a whole new cadre of depositors who look at the companies—including banks—in a different way. They feel that a bank that invests in small businesses, nonprofit organizations, and environment-friendly projects and causes, must be a good institution with which to do business. Banks operate under a Community Reinvestment Act, but even beyond these mandated requirements, some banks have found that socially responsible investments are simply good business.

Socially responsible banking

The Vermont National Bank (VNB), Brattleboro, Vermont is not the largest bank, but it is the best-known fiscal institution that has made major efforts to develop socially responsible banking (SRB). They formed an SRB Fund and, within a couple of years, attracted tens of millions of dollars in investment funds, much of it in new, out-of-state accounts.

"Joining Vermont National Bank in its concern for people, places, and the economic and social relationships on which they depend," stated a letter from Vermont National Bank, "is an admirable step in making our communities a better place to live." Numerous other banks throughout America, having heard of Vermont National Bank's success, have inquired about the SRB Fund. The *New York Times*, reporting on Vermont National Bank's environmentally friendly approach to investment, headlined the story "Bank Puts Your Money Where Your Heart Is." That's the type of public relations that's hard to buy.

Socially responsible investing

The growing trend away from traditional investment methods—using profitability as the sole criterion—still has to be proven on a large-scale basis. There are enough indicators, however, to allow some investors and traders to blend their business motives with environmental consciences. Socially responsible investment firms are not your ordinary Wall Street lions. They are out for profits, of course, but they have seen the profit light in issues that can also clean up existing pollution and make new investments in companies that have found ways of keeping the old globe livable.

Mutual funds

Jerome Dodson, Harvard MBA '71, is a ticker tape factotum who believes that taking social factors into account actually improves a mutual fund's performance. As he put it, "A company that treats its employees well has more internal harmony and can attract and retain more talented people who will work more productively."

In Dodson's opinion, which is often inconsistent with usual Wall Street opinions, environmentally sound, happy companies offer more financial stability, as well as promoting socially beneficial investments.

Early in 1989, there was only one fund that focused on environmental investments, the SFT Environmental Awareness Fund. Within a year, the investor had several choices in addition to SFT—Alliance Global Environment, Environmental Appreciation Fund, Fidelity Select Environment, Freedom Environmental Fund, Kemper Environmental Services, Oppenheimer Global Environment, and Progressive Environmental Fund.

Venture capitalists

Among the most cautious and shrewdest investors are the venture capitalists. This group pools their clients' money, as well as their own, and invests it in promising companies—usually at returns of 20 to 30 percent. Some of these venturers—sometimes called vultures—are introducing a social agenda into their investment practices, mixing entrepreneurship with social, health-care, and environmental problems.

The founder of one of the smaller venture funds, doing about $25 million in business, stated that, "A lot of social issues can be dealt with more efficiently by using the profit motive. What we are attempting to do is to hold together two seemingly, but not necessarily, contradictory principles: that you can do well financially and do good socially."

Another new development is the Social Venture Network (SVN). This is a loosely affiliated group of more than 200 venture capital groups that encourages the development of socially responsible businesses and attempts to steer the forces of finance into environment-friendly directions.

Wayne Silby, a Washington member of the SVN, stated that, "Many entrepreneurs don't want just a partner who is going to take more and more from them. We are good partners for people who want to preserve a certain amount of integrity." The group with which Silby and his partners John Guffey, Michael Tang, and John May are affiliated is called Calvert Social Venture Partners and is located in Bethesda, Maryland. But what is meant by a company providing social benefits? The following is a statement from Calvert's prospectus, which details the Fund's mission:

"We believe that by providing financial and managerial support to young, socially responsible companies, a new generation of American business can be created—one that realizes the vision of simultaneously creating economic and social gains. Our mission is to be a model institutional investor and innovator in the financial services field and, as such, we will only invest in those companies that demonstrate clear market and financial potential in areas that make positive contributions to the public good . . ."

While investment groups that have caught the torch of environmentalism are located coast-to-coast, another typical one in Vermont is The Catalyst Group, an investment group of "angels" founded by Bob Barton. This Brattleboro investment and banking firm was established in 1987 and concentrates on the $5 to $15 million company market that must include those involved in "environmental, socially responsive, and nonprofit businesses."

In San Francisco, another investment company is immersed in the Green Revolution. Hambrecht & Quist's Environmental Technology Fund has amassed $17 million worth of capital since its July 1989 creation. They received about 300 business plans during their first year and invested in three deals—companies involved in solvent and plastic recycling. Having environmentally friendly products or services is not enough, said H&Q. Applicants still need a sharp business plan and competent management to attract investment money.

Stock opportunities

If one problem can be perceived with investing in environmental stock issues, it is that the big investors got there first and skimmed some of the cream off the top early on. That's where smart money management comes in.

One example of such a stock market comet is Waste Management Inc., whose stock can be found in practically all environmental fund portfolios. When this stock hit the market in 1985, it produced a reaction not unlike a spray of water on a hot griddle. Within a short time, it rose 586 percent and even during the 1989–90 period, Waste Management stock climbed an additional 59 percent—selling at 25 times its estimated earnings for 1990. This is nearly double the price-to-earnings ratio of the stock market as a whole. Even when the Iraq oil crisis hit the headlines, these environmentally attuned stocks floated comfortably above their mid-range.

Nobody knows how long it will take—how many months or years—until we can clean up the environment befouled during the past century or so, but one thing appears clear: the companies that are in the cleanup business are cleaning up in the stock market.

The very Clean Air Act will be worth, by the time it becomes fully effective in 2005, an estimated $21.5 billion. Any company that is, today, in the environmental cleanup business or plans to get into it soon, must find this figure a mouth-watering incentive. No wonder investments are fattening if you own a fleet of garbage trucks, sell industrial air scrubbers, or manufacture and install water purifiers!

Socially responsible investment groups

The following investment groups offer counseling and asset management for those seeking profitable strategies consistent with their ethical beliefs and social ideals:

Progressive Securities
5200 SW Macadam Street
Portland, OR 97201

North Country Cooperative Development Fund
2129A Riverside Avenue
Minneapolis, MN 55454
(focusing on co-op groups)

Ethical Investments
430 First Avenue, North
Minneapolis, MN 55401

Social Responsibility Investment Group
127 Peachtree St., NE
Atlanta, GA 30303

Social Investment Forum
430 First Avenue
Minneapolis, MN 55401

Franklin Research & Development Corp.
711 Atlantic Avenue
Boston, MA 02111

Clean Yield Asset Management
224 State Street
Portsmouth, NH 03801

Robert Berend, JD, Southern California
Socially Responsible Investment
Professionals
210 S. Hamilton Drive
Beverly Hills, CA 90211

Affirmative Investments
129 South Street
Boston, MA 02111

Dreyfus Third Century Fund
767 Fifth Avenue
New York, NY 10153

New Alternatives Fund
295 Northern Boulevard
Great Neck, NY 11021

Paramus Fund
1427 Shrader Street
San Francisco, CA 94117

Pax World Fund
224 State Street
Portsmouth, NH 03801

Socially Responsible Banking Fund
Vermont National Bank
P.O. Box 804
Brattleboro, VT 05301

Working Assets Money Fund
230 California Street
San Francisco, CA 94111

Information sources

The *Directory of Environmental Investing* profiles more than 80 publicly traded environmental service companies. These profiles include vital financial statistics, technological capabilities, and potential for growth. Another 50 *Fortune 500* companies with significant stakes in the environment and the processes from which they will benefit are also reviewed. Interesting additional points covered in the 260-page directory are environmental laws and their costs, as well as trends in the environmental field. The directory can be obtained from Business Publishers Inc., 951 Pershing Drive, Silver Spring, MD 20910-4464 for $75. Ask for the newest third edition.

More and more publications are featuring socially responsible and environment-friendly investment avenues. Just a few selected ones include the following:

Clean Yield Newsletter
Box 1880
Greensboro Bend, VT 05842
Monthly newsletter on ethical investments/$75 a year.

Insight
711 Atlantic Avenue
Boston, MA 02111
Monthly investment advisory service on
environmental impact companies/$87.50 a year.

Good Money
P.O. Box 363
Worcester, VT 05682
Bimonthly newsletter on publicly traded
company that emphasizes social responsibilities/$75 a year.

News for Investors
1755 Massachusetts Avenue, NW
Washington, DC 20036
Nonprofit, monthly newsletter by Investor
Responsibility Research Center. Varying contributions.

Wall Street Green Review
24681 Alicia Parkway
Laguna Hills, CA 92653
Monthly publication by Environmental Investment
Strategies. Topics: waste services and alternative resources
industries/$48 a year.

Appendix B also contains a list of companies the environmentally caring investor might wish to explore.

6

Going abroad: exporters with an environmental conscience

EUROPE AND EVEN CANADA ARE WAY AHEAD OF THE UNITED STATES IN environmental entrepreneuring. Not that this came overnight! In West Germany*, a prominent product was first introduced in 1978 as an "environment-friendly" consumer item. It took seven years before sales grew sufficiently to generate corporate profits. Today, it is one of the leaders. Everything has its time and place.

American companies with heavy financial interests in Europe and lots of experience are correctly advised by Procter & Gamble's director of environmental coordination who said, "European consumers are five to seven years ahead of the United States in their environmental concerns. Even Canadian consumers are a few years ahead of the United States."

Those environmental Europeans

Larger, more industrialized countries in Europe have proportionately larger problems. However, some aspects of analyzing the European

*Although the problems of "East" and "West" Germany will now have to be addressed by the newly united country, for purposes of this discussion, the old designations will be used.

Community consumer market remains fairly constant—every day is Earth Day. Europeans are not as spoiled; they do not waste as much as we do. Take the life of the average German *Hausfrau*.

She gets up in the morning to take a quick shower. Since hot water costs money, she turns it off while soaping, then on again to rinse. Rather than turning up the heat too much, she slips on warm clothes. Breakfast is completed with piling empty deposit bottles into a reusable cloth or plastic carrier. She puts them into the rear basket of her bicycle and rides down to the grocery store to exchange them for about 18 cents each. When she returns, her bike goes back into the garage or storage area before she walks to the bus stop for a ride to work or into town for shopping. In smaller towns, she would ride her bike into town or to work, just as thousands of others are doing daily.

Europeans have learned to bi-cycle and to re-cycle. They are aware of environmental issues. They have discovered that it is possible to be both environmentally conscious and economically prosperous. The compactness of the smaller countries of Europe has tended to accelerate the progress of environmentalism and the adjustments that entrepreneurship had to make on the way to the European pocketbook. Americans who hope to do business abroad need to understand this. U.S. exporters will find in Europe a ready market for environment-friendly products and services—however, not without considerable native competition, some of it already in place for a decade.

The geo-political movement, and in Europe, environmentalism is closely tied to politics, has broadened opportunities for American business. It is pronounced in virtually every West European country and is now making giant inroads in the East bloc countries as well.

Marketing American products abroad will demand greater efficiency in production, not necessarily higher prices. Some increases can be absorbed but most of our higher prices can be contained by following the examples of our overseas customers. West Germany creates half as much solid waste per capita as we do and creates 40 percent less carbon dioxide. Japan now uses one-third as much energy per capita as the United States. During a past decade of enormous growth, Japan actually used 40 percent less energy and raw materials in production than we did.

Case history: Switzerland

Not all efficiencies observed overseas are due to efficiency in production. Some are due to the very compactness of these countries, as well as the homogeneity of their populations. Take, as an example, little Switzerland.

Any traveler who has traversed even a portion of this whistle-clean little nation of a mere 16,000 square miles, home to around 6.5 million people (1/225th the size of the U.S.A.), must have been astounded at its pristine and nearly sanitary condition, its scrubbed natives and neat cows. Still, the Swiss, in the final decade of this century, have a pollution problem of substantial proportions.

Starting around 1975, the Swiss began one of the most advanced environmental management systems in the world. The Swiss Federal Office for the Protection of the Environment is, of course, helped by a blessed topography that is virtually self-cleaning and a people who are just naturally neat and conserving.

With all the natural help and compactness, the Swiss still needed a combination of political will, legislation, massive application of available technology, state-of-the-art waste management systems, and a countrywide system of local coordination. Today, 84 percent of all water passes through central purification stations. Manufacturers are required by law to perform primary treatment of their wastes on-site. Licensed operators dispose hazardous wastes in high-temperature incinerators, including dioxin. High-phosphate detergents were outlawed. Finally, all communities were obliged to improve their current treatment plants.

The federal government's role was that of a promulgator of regulations, coordinating them with local subdivisions, and underwriting from 30 to 40 percent of local costs—about 30 billion Swiss francs to-date. While the Swiss cleanup began as a very relaxed, benign, and democratic process, even here, laws and penalties became progressively more stringent. The task was too formidable for local municipalities to execute; growth in population and concomitant pollution too overwhelming. Private industry was called upon to lend a hand, and businesses responded with investments of an additional 7 billion Swiss francs.

The key to the success of the Swiss environmental cleanup and future control has been the high degree of cooperation between all levels of the government. As a Swiss environmental official put it, ". . . it is a problem of politics rather than techniques. The technology has been known for years, but if each little village makes its own decisions, then implementation is impossible."

There is one other aspect of the Swiss's lack of pollution. Keeping the environment clean starts with the individual. Consumer education, in Switzerland, as in the United States, is a necessary ingredient in any eco-entrepreneurial campaign.

Case history: West Germany

West Germany is second to the United States in the amount of waste it accumulates. Civilization and progress appear to go hand-in-hand. There is one difference in Germany, however. It does not have the space that the United States has to bury its refuse. While some of Germany's landfill contributions are being buried in contracted French and former East German facilities—where more open spaces exist at this time—that expediency is also gradually being used up. Hence, new and more innovative methods of refuse and garbage disposal had to be found. The following is a random list of methods that have been, and are, being employed in West Germany. Many of these methods are implemented by private contractors; others are being privatized gradually as they are often more economical. No matter, all of them can be inspirations to parallel entrepreneurial activity in the United States.

- Federal seals of approval—called Blue Angel—are placed on products approved as environmentally friendly by a jury of peers."
- Unused medication is returned to pharmacists for disposal.
- Graduated scale of waste disposal fees—the less you dispose, the lower the per-pound rate goes.
- Lower collection rates for households that separate wet garbage from recyclable product wastes.
- Issuing only one can per household for co-mingled wastes, with smaller fees for smaller containers.
- Two-can system for compostable wastes and remaining refuse.
- Banning of compostable wastes from some landfills.
- Collection trucks equipped with printout scales that weigh garbage and charge accordingly.
- Numbered stickers from 1 to 52 sold to each household at start of year. Any stickers left over at end of year will be redeemed for cash or credit. Householder or office attaches one sticker in numerical sequence to each waste receptacle.
- Backyard compost bins available to householders for a small fee.
- Educational video used in stores extolling the use of environmentally friendly "Blue Angel" products.
- Imprinted cloth shopping bags given out by grocery stores, or sold at cost, and subsidized by the government.
- Coordinated educational programs in conjunction with stores, ad media, schools, and commercial and industrial programs.
- Neighborhood resource recovery bins, often with a density of one bin per 1,000 homes, that collect newspaper, aluminum cans, et. al. These are often joint ventures with local governments.

Export product cautions

Americans planning to export U.S.-made products to the European community will need to observe some environmental caveats. Not that all products used by Europeans are environmentally safe or friendly. Far from it. But the future, and to a large extent the present, will make marketing there a lot easier if you consider the following, existing limitations:

- Aerosols—being phased out everywhere. No CFCs, no butane. Switch to pump-action containers. If they are re-usable, so much the better.
- Diapers—disposables or "nappies" (in Great Britain) are no longer bleached with chlorine products. Procter & Gamble, Peaudouce, and a Swedish company are already heavily represented with environment-friendly products.
- Detergents—many restrictions, especially on the use of phosphates. Eliminate the latter altogether. Restrictions written into local laws exist in Switzerland, Germany, Denmark, Sweden, Netherlands, Italy, and parts of Great Britain.
- Pricing—some environmentally safe products are slightly higher priced, but for the most part, companies have maintained comparative price structures. They depend on production efficiencies, lower packaging costs, and increasing demand.
- Automobiles—The European community has imposed U.S.-style standards on the industry, effective 1993. The impact will be enormous. Since three-way catalytic converters (to screen out hydrocarbons, carbon monoxide, and nitrogen oxide) work only with fuel-injection systems, 70 percent of European cars not now made with them will have to be retooled. This means good business for producers of auto emissions equipment.
- Soft drinks—Already Pepsi Cola has moved into Russia with a well-publicized presence. Keep in mind, however, that the average European drinks only 15 gallons of carbonated soft drinks a year, compared with about 50 gallons for the average Yankee. Soft drinks, however, are very price-elastic. As of 1990, convenience packaging—plastic or aluminum and tin cans—dominated 90 percent of the European market. However, many Europeans are used to glass deposit containers, and aluminum is 100 percent recyclable. Its scrap value is high, especially in Sweden, which recycles more than 80 percent of its aluminum containers. Other countries are following suit or passing laws that impose heavy deposits on plastic containers, or even restricting their use to a small proportion, such as in West Germany. Denmark has banned disposable containers altogether.

Eastern Europe's problems

"The situation in the emerging democracies of East Germany, Poland, Czechoslovakia and Hungary is something like environmental bankruptcy," said Klaus Toepfer, West German Minister for Environment.

Using East Germany's own numbers, 30 percent of its bodies of water are ecologically dead; 25 percent can no longer be used for drinking water even if treated by modern technological methods; and factories spew pollution as they burn the region's available high-sulphur lignite coal.

The Poles, perhaps more than their neighbors, are doing as much as their meager resources allow. Solidarity, the major labor movement now virtually running the Polish government, has supported the Polish Ecological Club and, in turn, is advising the government on environmental matters. The U.S. spokesman, Zbigniew Bochniarz, announced that a series of "green libraries" will be established that welcomes contributions from abroad. The first two in operation are located in Poznan, Poland and in Riga, Latvia.

A green revolution among the reds

The Soviet Union is one of the richest states on earth if judged by its natural resources. For the first 70 years of the Communist regime, the philosophy "take, take it all" prevailed. Since Mikhail Gorbachev's ascendancy, Russia has begun to join Western nations in their concern for the environment.

All of a sudden, the ministers and achievement-driven factory managers have discovered that they are breathing the same air and drinking the same water as the ordinary worker. Soviet ministries, comparable to Western conglomerates and major corporations, were only concerned with quotas and profits. Until now, that is.

Case history: Kuybashev

In the Soviet city of Kuybashev, lying in the Volga basin 600 miles east of Moscow, nearly 1.5 million people now have another police watching them. The new 70-member force is the Public Health and Ecological Militia (PHEM). Kuybashev, the wartime Russian capital when the Nazis were knocking on Moscow's door, has a giant hydroelectric plant, many other polluting industries, and countless trucks rumbling daily in and out of this industrial and grain center. The PHEM police will now be keeping an eye on how enterprises, institutions, cooperatives, and individual dwellings observe recently established environmental health standards. Lawns, parks, squares, and open spaces are also in the new police force's domain.

There is little doubt that Soviet citizens in and around Kuybashev will have to learn about cars and trucks with defective exhausts and get them repaired pronto—or face fines and sanctions. Fines range from 10 to 50 rubles for violations to local health standards. It is quite likely that by the time you read this, other polluted Soviet cities have followed suit.

Careful cooperation

Not long ago, a conference of environmentally concerned Russians took place to assess the actual ecological situation in their homeland. One of them was a communist delegate from Tallinn, Estonia, named Juhan Aare, who is also a leader of Estonia's Green Movement. Aare made the following announcement at the conference:

"People of the United States and the Soviet Union share the important task of cooperating to save the environment. It unites us. Your Western big, multinational corporations are interested above all in earning greater profits and are not so concerned with saving the environment. But on the other hand, the United States has the press and a political system that can effectively influence those conglomerates. I think Soviet ministries and U.S. firms resemble each other in wanting profits. But I think the firms' presidents and representatives understand that a person can be very rich, but that without clean water and air, wealth has no sense . . . We must admit that the situation is not just critical, but catastrophic . . . It is clear that without international cooperation nothing can be achieved . . . If you are willing to cooperate with us, we also are ready."

The Greenpeace office of Moscow, headed by Olga Lysenko, issued the following caution: "We're concerned that Western companies will look at us as a market for pesticides and technologies banned in the rest of the world." This statement indicates a great deal of sophistication and awareness of world affairs.

Such statements by highly placed Soviet officials appear to indicate that American eco-entrepreneurs have an opportunity to do business in Russia as well as in other ex-satellite countries. But we need to keep in mind that similar standards, as in the United States, must be observed.

Currency problems in Eastern Europe

One problem that needs to be addressed in Eastern Europe is hard currency. One way to overcome the lack of hard currency of potential East European buyers is to look to our government and its overseas loan program. The Environmental Protection Agency in Washington is starting a program in Hungary that might be the forerunner of many other programs. It is spearheaded by specialists in Washington and funds from the U.S. State Department's A.I.D. coffers. (See chapter 9 for more on this program.)

The Swedes have taken what appears to be the first step in realistically appraising the crumbling communist economies. The Swedes, as well as their neighbors in Denmark and Finland, have joined in allocating

$150,000 for an independent environmental group in Poznan, Poland. The latter will establish a network of water-quality testing stations along Poland's Vistula River. This river flows into the Baltic, carrying pollutants with it into Scandinavian fishing grounds.

In addition, Sweden has pledged another $50 million to another joint cleanup effort operated by a Polish-Swedish group, including wastewater treatment facilities along the Vistula and the first major conversion of coal-fired furnaces to natural gas systems in the entire city of Krakow (population 750,000).

Other Western countries like Germany, the Netherlands, and Finland, are planning similar programs with Poland and other Eastern countries.

Opportunities for Western entrepreneurs

With the problems of a thoroughly polluted and dangerous environment staring them in the face, Eastern Europe and Russia seem ready to do *something*—even if this means accepting help from Western countries. Consequently, air and water pollution control devices and chemical, as well as proven technological skills, would seem to find a ready market in the vastness of Gorbachevland. Eastern countries, within the limits of local fiscal resources and American aid, provide untapped new opportunities for manufacturers of nuclear and natural gas (clean-energy) plants and industrial pollution-control devices. Here are just a few of the existing specific opportunities:

- Aral Sea, Russia—huge salt flats created by the drainage of this inland sea to water new cotton croplands adjacent to it. However, 40 percent of the formerly fish-rich sea has disappeared and occasional salt storms from the denuded and exposed shores are poisoning the very cotton lands the sea was intended to nourish.
- Prague, Czechoslovakia and Budapest, Hungary—recipients of the highest acidic rainfalls, made poisonous by unchecked smokestacks, have turned historic statues into blackened stumps, as the soiled, corrosive air eats away at them.
- Sudety Mountains, Poland—large patches of forest have been destroyed due to acid rains, polluted by nearby factories. Local researchers have estimated that by the end of this decade, 50 percent of Poland's 32,000 square miles of forest lands will be damaged if this situation is not controlled quickly.
- Rivers in Poland, Czechoslovakia, and Russia (primarily in the Volga basin)—much water has become unfit, even for filtration, and would corrode pipes. Prague admitted that 28 percent of its polluted rivers are devoid of fish. The South Volga River receives 5 trillion gallons of wastewater annually and double that amount

> is drawn out for agricultural and industrial use. The Volga is literally on the brink of disaster.

The new revolution in Russia is not merely a monochromatic red one, but has added *green* to its astounding planning. American eco-entrepreneurs seeking to do business there and throughout Eastern Europe may well prosper. The greening of Europe continues and American exporters will need to stay in step if they wish to reap the green.

7

Case studies: successful green entrepreneurs

IT IS MY HOPE THAT READING ABOUT HOW OTHER ENTREPRENEURS have turned their concerns about the environment into profitable businesses will set your own juices flowing. Along with the case histories presented in this chapter, there are some concrete ideas about creating actual business ventures that can help your pocketbook and your world. The ideas are presented in alphabetical order by industry.

Advertising: demarketing for socially positive results

Sophisticated techniques that sell billions of dollars worth of goods and services are now being used to put the brakes on sky's-the-limit marketing. Organizations are discovering that one can use commercial advertising to modify attitudes and behaviors, such as supporting environmental causes and altering indifference toward pollution.

Essentially a form of behavior modification, social marketing is regarded as noncontroversial and even preachy. It used to be called institutional advertising. Major corporations are jumping on the bandwagon with wordy ads extolling their environmental concerns and programs in prestigious major media. Sometimes, these messages are great; sometimes, they are a lonely bandleader waving an earnest baton before an empty orchestra—the headquarters folks are cognizant of the problem but the middle management people haven't heard the solutions.

Such advertising is being used to counter smoking, drug use, illiteracy, energy abuse, pollution, and waste. Today, ads extol moderation, responsible drinking, safe sex, and even mental illness. Sometimes, such social marketing is credible; sometimes it appears too pontifical to have any impact. At best, it contributes to awareness for a cause or against a form of antisocial or destructive behavior.

Responsible Corporate Advertising

"The direction of an environmentally conscious style is not to have conspicuous consumption written all over your attire. We believe this could best be achieved by simply asking yourself before you buy something (from us or any other company) whether this is something you really need. It could be you'll buy more or less from us, but only what you need. We'll be happy to adjust our business up or down accordingly, because we'll feel we're then contributing to a healthier attitude about consumption. We know this is heresy in a growth economy, but frankly, if this kind of thinking doesn't catch on quickly, we, like a plague of locusts, will devour all that's left of the planet."

(From an advertisement by Esprit Corporation, Spring 1990)

While many large ad agencies create such advertising for their *Fortune* 500 clients, and pro bono agencies like the Ad Council and others contribute more than $50 million worth of talent, space, and time, there is one small ad shop in California that is set up to produce environmental ads exclusively.

Right from the start, Stephen Garey decided not to be a town crier for capitalism, but to use his skills and talents, and those of his associates, "in the service of any firm whose product, service, and method of manufacturing brings no harm to our planet."

An under-50 defector from the East Coast, Garey gathered a few like-minded associates and decided that they wanted to run an advertising agency devoted to the environment. An advertisement in the *New York Times* produced remarkable results. It stated Garey's philosophy as follows:

"Despite overwhelming evidence of serious damage to the Earth, tens of thousands of companies, large and small, continue to make products and profits in ways that can only lead to further ill health and tragedy . . . Such businesses can no longer be viewed as successful . . ."

Considering that about 300 inquiries were received by the Garey group, there can be no doubt that an entrepreneurial market exists in the United States for marketers with a conscience.

Agriculture: feeding 10 billion Earthlings in 2092

The 20th century is known as the period when agricultural populations shrunk steadily. Millions of people, like lemmings, rushed into nearby cities for better education, economic opportunities, and the trappings of civilization.

On both sides, environmental deterioration took place that has begun to endanger the very survival of peripatetic populations. In the coming decade and beyond, agriculture will have three main areas of opportunities: (1) reducing reliance on pesticides and fertilizers; (2) controlling the demand for irrigation water; and (3) increasing crop production yield per acre.

For agriculture to satisfy the needs of a world population that could double in the next century—from the current 5 billion to a humongous 10 billion by 2092—no single improvement will be adequate. Entrepreneurial opportunities exist in such areas as grazing control, nontoxic fertilization, genetic improvements of both crops and food animals, soil erosion control, optimum water utilization, irrigation efficiency, alternate energy developments in remote areas, control of encroaching desert areas, desalination of soil and water, but most important, educating vast populations in new agriculture techniques and the preservation of the existing, fragile environment.

Beer sludge is farmlands' fudge

Nearly everything can be recycled with a little effort if it compounded with man's ingenuity. In Jacksonville, Florida and Fort Collins, Colorado the Anheuser-Busch Companies have two large breweries. In the process of producing beer, the remaining sediment of grains, hops, malt, and other by-products used to be considered waste. Now, the company has found that these leftovers can be reprocessed into a nutrient liquid and sold for fertilizer.

The beer sludge is mixed with some water and sawdust. It contains all organic ingredients rich in carbon, nitrogen, and trace nutrients. Each day, 50 tons of the sludge is produced, and instead of clogging up landfills as before, the mixture is now sold as a composted nutrient to farm operators and landscape companies.

Such ambitious and imaginative enterprises combine environmental concerns with entrepreneurial cleverness. There are nearly a 100 major breweries in the United States and many hundreds of small, localized ones. It would not take too much of a survey to determine whether this ecologically desirable business also makes economic sense.

Composting without aromas

On his farm off Route 109 between Barnesville and Comus, Virginia, Peter Knop turns large tree stumps and branches into compost. It is called the Ticonderoga Process and represents a new approach to dealing with yard and tree wastes that could help solve local landfill and waste disposal problems.

Usually, wood waste is shredded and mixed with other materials for composting. Knop's method takes the whole stumps, arranges them in long windrows up to 18 feet in height, and allows them to decompose naturally over a period of four to seven years.

The farm takes yard and tree wastes from others for $29 a ton—which compares favorably with the county's current fee of $53 a ton. Residential tree and yard wastes from neighbors are accepted at no charge.

In the Ticonderoga composting process, plant wastes are "inoculated" with saprophytic fungi, the type that grows on dead wood. The windrows are then planted with wisteria, ivy, and other vines and blackberries. The latter send their roots into the wood and act as a vehicle for the fungi and oxygen to carry out decomposition.

The vine cover acts as an insulator to keep moisture and temperature levels constant, very much like a forest canopy would. The Knop system is plant-based versus bacteria-based. It does not have the odors associated with traditional composting operations.

Knop uses the composted humus to improve the soil and found that his nursery plants and pharmaceutical herbs grown on the farm for resale, did much better. Former expenses for pesticides and fertilizers have virtually been eliminated since using the new composting material.

In 1989, the farm took in 150,000 tons of debris for composting between April and December alone. The Barnesville, Virginia farm has come to the attention of the state of Virginia, which is using this site as a pilot program.

Herb gardening: add spice to your life

More and more popular is herb gardening. Small farms, suburban acreage, and even backyards are ideal for pesticide-free herb farming, which can produce high-yield, high-profit crops. According to the American Herb Association in Rescue, California, "natural food continues to grow in popularity . . . Markets for herbs will grow, and with it, a demand for additional growers."

A billion dollar herb market exists in the United States if one counts the "biggies" such as Herbalife, Celestial Seasonings, Traditional Teas, and Spice Island. But small entrepreneurs can get into the business on a part-time basis for an average upfront investment of about $3,000. Current experience indicates that an investment of $30,000 overall can return about $100,000 gross and a pretax profit of about 40 to 45 percent.

In herb farming, the greatest single start-up cost is land. Herb Nursery of Vista, California, run by Kent Taylor, started out with five acres but has now expanded to 25 acres. The company currently ships live, multiple-set plants all over the country, selling as many as 500 orders a week through their catalog.

In Kansas City, Missouri, Herb Gathering, a cooperative of 10 growers was formed to satisfy the fast-growing herb market.

The restaurant market is also a lucrative one for herb growers who can supply reliably in sufficient quantities. It is probably better, therefore, to grow a few herbs in quantity rather than a large variety of them—adding new varieties as the market warrants.

Some herb growers have also taken advantage of the growing interest in herbal medicine. Natural food shops and some "natural" medicinal manufacturers rely on dependable small growers who can supply them in bulk.

Rent-a-tree for city slickers

In numerous suburban communities, developers and home owners associations have set aside small plots of land on which local residents can try out their green thumbs. The inspiration, of course, came from World War II's Victory Gardens. Now, an enterprising company in Cambridge, Massachusetts has come up with a way to rent a bit of nature to city dwellers who hanker for the country. North Country Corporation is offering nonfarmers an apple tree, a maple tree, or other trees, and even a beehive, which they harvest and ship to the lessor at an annual fee of $37 to $39. Leases can be assigned and given as corporate gifts, for example, and come fall, lease owners can taste their own right-off-the-tree MacIntoshes or slather fresh maple syrup on their pancakes.

North Country guarantees to ship renters at least a half bushel of apples—about 17 pounds. More than a thousand people have taken the company up on their offer and some 5,000 have purchased the proceeds of their leased maple trees. The company embellishes the promotions with educational and often amusing progress reports, hints on the harvesting process, and even recipes for making applesauce and maple candy.

Making Mother Nature your partner in an enterprise that benefits thousands of customers is a perfectly ingenious example of how the environment and entrepreneurship can work together.

Synthetic (safer!) pesticides

According to the ICI Chemical people, the sawtoothed grain beetle is responsible for a $20 billion annual crop loss. Without the use of pesticides to fight this voracious insect, far greater agricultural losses would be sustained, says the company—food that now feeds nearly 2 billion people around the globe.

Pesticides have their drawbacks, however, and that's why the latest biochemical research is important for entrepreneurs in agriculture. Here is what ICI stated in one of their announcements, which was published in *The Wall Street Journal*:

"We've developed synthetic pyrethroid insecticides, which mimic the chemistry and environmental qualities of chrysanthemums. Pyrethroids tend to be more effective and work at lower concentrations than other crop protecting chemicals."

While the above research into synthetic pesticides goes on, biotechnicians are also at work trying to develop high-yield, disease-resistant crops that will naturally require fewer additives of any type.

Topsoil producer takes 500-year shortcut

Five hundred years since the "discovery" of the American continent by European adventurers, nearly 500 million people in the Americas have depleted much of the good topsoil that originally covered these lands. During the same time, new topsoil has been created, because it takes just about that long for nature to replenish the soil.

Not everywhere does man give the earth a chance to replenish itself naturally. That is where the Kellogg Supply Company of Southern California comes in. They "manufacture" topsoil. It has been a 30-year process and it has been so successful, both as a business and as an environmentally friendly contribution, that the Sierra Club and the California Environmental Council have lauded Kellogg's. It is said that in Southern California, Kellogg Supply has helped to keep the Pacific Ocean cleaner.

The company, mostly family operated, cleans and blends 400 tons of wet sludge and 1,000 cubic yards of wood waste each day and keeps nearly a million pounds of wastes out of the sewers and landfills daily. The resultant blend or "soil amendment" increases crop yields and acts as a bridge between waste disposal and garden industry supplier.

Kellogg's is the perfect example of the recycler who has blended the needs of an industry with maintenance of the environment—and to the profit of both.

Aquaculture: a good catch for entrepreneurs

The more than 500 established fish hatcheries are going to get lots of competition in the years ahead. Aquaculture, the raising of fish in man-made environments and under scientifically controlled conditions, is a growing business for entrepreneurial specialists as well as for farmers. The latter often raise animals anyway and are use to this activity. They often have a barn or other indoor facility where a modern, scientific aquaculture installation can be accommodated. And they very often have a marketing-distribution system already in place.

One company that does a sizable business in indoor fish farming is Fresh Culture Systems, Inc. of Kutztown, Pennsylvania, run by Richard

Fahs and Steve Van Gorder. For about $165,000, they will install a 15-tank system that can produce as much as 50,000 pounds of fresh fish annually. The way fish is selling today, this means that the basic investment can be recaptured in a little over a year. Sometimes, technological help can be obtained from a nearby university. In the case of Fresh Culture Systems, the Ben Franklin Center of Bethlehem, Pennsylvania's Lehigh University, provided much of the skilled assistance needed.

That aquaculture is a growth industry can be proven by the U.S. Department of Agriculture's figures for 1980—when 203 million pounds of fish were raised. Today, that figure is closer to 1 billion pounds. While most of today's aquaculture is being developed in the more benign climates of southern states, a good, tight, 5,000-square foot building elsewhere in the country will do just fine. Just remember that fish don't swim in ice.

Fish business going up in smoke

For Karen Ransom and Peter Heineman, $6 million worth of fish are literally going up in smoke. Their company, Homarus Inc. of Mt. Kisco, New York takes tons of raw fish, spices them, smokes them, and then sells them to the most prominent chefs in New York and nearby areas for somewhat elevated prices.

This super-delicious and environmentally friendly business—unless you are a pure vegetarian—was started in 1976 with little more than an idea, a resolve, and a series of lucky coincidences.

Heineman participated in a college work study program in California that involved aquaculture. Some sound advice from his father and a concomitant visit to an Idaho trout farm convinced him and his new bride that a superior smoked fish product could be a good entry into entrepreneurism.

What clinched the theory into solid reality was a further visit to a friend in New York who was a chef in a famous restaurant. The latter tasted the smoked trout that Peter and Karen had prepared in a small smoker given to them by Peter's mother—and, as they say, the chef flipped. He asked the Heinemans to make him some more and even recommended a number of other Manhattan chefs as further contacts. We could say that the rest is history but that is usually not the way entrepreneurism works. Cleaning, curing, and smoking nearly 1.5 million pounds of gourmet fish to be distributed and shipped immediately is w-o-r-k.

During the past decade, Americans have become acutely aware of proper nutrition and the value of fresh food. At the same time, the American palate has become more sophisticated. A solid U.S. economy has also made such gourmet products as smoked fish more affordable, all of which makes for a robust future for those entrepreneurs involved in aquaculture and aquaculture equipment, as well as hydroponics.

Hydroponically grown produce can be grown inexpensively and more environmentally sound than conventional farming methods. It takes less soil, wastes little water, needs minimum space, and makes pesticides virtually unnecessary. And as with aquaculture, entrepreneurs can grow year round and in any climate.

It is no wonder then that aquaculture and hyproponics offer virtually limitless potential for entrepreneurs who recognize their valuable environmental contribution as well as their ability to meet today's growing consumer's demand for fresh, nutritional food.

Architecture: blueprints for healthy designs (see also *Homebuilders market* and *Remodeling*)

Designing homes and offices that are environmentally safe and harmonious with nature is William McDonough's calling. The New York architect is right in tune with the American Institute of Architects, whose spokesman, Doug Greenwood said, "Architects are perhaps more aware than most of the effects they can have on the environment." Concern over ecological issues is becoming a mainstay in modern architecture.

McDonough's philosophy is that, "Things cannot be beautiful if they don't work for people and for the planet." That means using as many recycled materials as possible; making sure that plenty of fresh, clean air circulates throughout each building to cool it naturally and to vent toxins; avoiding woods that tend to deplete endangered rain forests (like mahogany); and using the most energy-efficient appliances and lamps.

For office structures, McDonough tries to make the indoors as much as the outdoors as possible. He says that people will be happier working in such an ambiance. For residences, he aims for high ceilings, tall windows, natural light, and proximity to broad-leafed trees that help to block the sun.

Until his own ecological building newsletter comes out, William McDonough recommends reading *The Natural House Book* by David Pearson (Fireside, 17.95) and *Your Home, Your Health and Well-Being* by Rousseau-Rea-Enwright (Ten Speed, $19.95).

Automotive opportunities: full speed ahead

The future of the automotive industry and the entrepreneurial opportunities are tremendous—no matter whether we find the next oil gusher, switch to electric cars, or find some other substitute for oil. One thing that must go on is automotive repairs.

Repair with environmental care

There is no doubt a few very conscientious auto repair shops already do an "environmental" job on your car. Jeff Shumway, whose Ecotech

Autoworks is described in the Introduction, has this to say about automotive repair:

- Leaking air conditioners and environmentally ignorant service procedures are the number one cause of upper atmosphere ozone depletion in the world. We collect and recycle A/C gas (Freon) on-site and detect/repair all leaks.
- Antifreeze is a serious, long-term surface and groundwater pollutant. We rejuvenate and recycle antifreeze on-site. Nothing goes down the drain.
- Conventional auto repair utilizes extremely hazardous materials. We use all available, less-toxic and nontoxic parts and chemicals—i.e., non-asbestos brake pads, eco-safe cleaners, *elbow* grease, etc.
- Recycling is only efficient as end-product demand. We offer guaranteed remanufactured parts, as available, on request.
- We participate in all available off-site recycling programs: used oil, scrap metal, dead batteries, rebuildable parts, office paper, cardboard, etc.

Clearly, his shop demonstrates that ecological concerns must start at the top, that consumers will respond to logical and reasonable education and that environmental concerns can be profitable, too.

But what happens to old, worn out cars? Are they destined to slowly rust in unsightly auto graveyards? To continue spewing forth dangerous emissions? Unocal doesn't think so.

Gas station owners/managers who want to do their customers a favor by protecting their health, as well as keep them coming back, might want to post or otherwise disseminate these Occupational Safety and Health Administration warnings:

- If you are going to pump your own gas, stand upwind from it so that you won't inhale the toxic benzene fumes from it.
- Keep your car windows rolled up while pumping gas so that fumes are prevented from forming inside the vehicle, especially if you have children in the car.
- Should any of the gas you pump spill on your skin, wash it off right away. It is toxic.

Gas stations: a geiger in the underground tank

There are more than 130,000 gas stations in the United States. Add to these the many marinas, large farms, corporate service facilities, and transportation companies, as well as military installations, and figuring that each underground storage ''field'' has one to four tanks buried in the earth, and you can come up with a realistic figure of about 400,000 plus tanks. These could be environmental time bombs ticking away.

In recent years, the Environmental Protection Agency has become cognizant of the real and potential problem of underground gas tanks. Tough insurance demands, in case one of these tanks springs an underground leak and befouls an aquifer carrying drinking water, have created substantial fiscal and political problems. But they cannot be ignored.

Gas station owners, especially independent ones who are not covered by large corporate insurance umbrellas, are caught in a catch-22 situation. Like doctors who are faced with insurmountable liability insurance premiums, smaller gas station owners need to have adequate insurance in case of an ecological disaster. The fly in the ointment is that insurance companies have raised premiums to such high levels that they have become unaffordable or imposed such high deductibles that small station owners would be put out of business if they were to be faced with a $10,000 to $20,000 underground test, or a $40,000 or more tank upgrade.

Today, 43 states have created insurance funds, similar to those available to savings and loans and banks. In some cases, the mandated deductible clause is too high. In other cases, insurance companies have simply dropped ''pollution'' from their coverage. While many larger stations will manage with self-insurance, the small, independent gas station owners might become a dying breed.

Is there an opportunity for eco-entrepreneurs? Gas stations along roads, shores, company yards, and depots will go on. Those entrepreneurs who know how to test for underground leaks, repair and replace faulty underground storage tanks—especially replacing old metal ones with noncorroding fiberglass or plastic ones—can look forward to many boom years.

Meanwhile, keep tuned to your state legislature and the insurance companies for further developments.

Used cars put to good use

Los Angeles has admittedly one of the nation's worst air pollution problems. For an out-of-hand cost of about $5 million, a local petroleum company called Unocal is willing to put a real dent in the problem. Uno-

According to EPA estimates, the American public drives 3 billion miles each and every day and uses up 82 million gallons of gas. This 82 million gallons of gas, processed through carburetors and engines and exhausted through tail pipes into the very atmosphere we breathe, adds about 1.5 billion pounds of carbon dioxide to the atmosphere—just on one single day in the United States.

cal will buy up all pre-1971 cars that are registered in the area for a flat $700 a piece.

The reason that pre-1971 vehicles are targeted is that these smoke-belching monsters emit 15 to 30 times the exhaust that today's cars emit. Clearly, this type of program will help reduce the infamous L.A. smog.

The program is called SCRAP, an acronym for South Coast Recycled Auto Program. The old cars will be turned over to a scrap yard that will dismantle them, sell off any usable parts, and recycle the rest of the rusty wrecks.

The value of a used car in terms of recyclable parts and metal is easily determinable. Perhaps, if enough used car dealers across the country got into the act, several good things could happen: (1) the environment would be cleared of smokey clunkers; (2) used car dealers would acquire a powerful public relations tool; and (3) a market would be created for replacement cars, both new and used.

Batteries with nine (or more) lives

Americans use up nearly 2.5 billion batteries each year. In some homes, equipped with gadgets, portable radios, remote-control toys and portable computers, $50 to $100 a year goes into the dump in used up batteries. Disposable batteries are not only an economic waste, but also a landfill hazard. They contain mercury, cadmium, and other toxic metals that seemingly cannot be recycled.

There are battery rechargers and batteries that can last a lifetime, however. They are made by Gates Energy Products, Gainesville, Florida, which has actually been manufacturing them for 25 years but only for other battery companies. Gates' Millennium batteries carry a price tag of about $6 per pack. A RapidCharger that rejuvenates four batteries at a time costs another $20, and they are now available over the retail counter.

Eveready Battery also is marketing a recharger called the Generator. It also recharges four batteries at a time, but in just one hour, whereas Gates' charger takes three hours. Eveready's price is $49. Panasonic is another recent entry into the charger field with its SlimCharger, a $27 seller that takes six hours to revive four batteries.

Beauty products retailing: in the pink

There is no doubt that many beauty products shops sell products to enhance milady's looks, but there is only one *The Body Shop*. Spawned in 1976, by Anita Ruddick in England, and starting with a single shop capitalized at $6,400, *The Body Shop* now grosses $141 million and shows a pretax profit of $23 million. What makes *The Body Shop* so bloody special is that they sell ecology-conscious cosmetics.

Ruddick's shops never use animals for experimentation of their cosmetic products. They use only ingredients that have been used by human beings for centuries and are entirely natural. Each product explains how and why that product was created; each label is a consumer education tool.

Ideas and ingredients come from all four corners of the earth, which puts this entrepreneur on the road for a good part of the year. Information is passed on to consumers and they become part of Ruddick's classroom (she is a former history teacher). Not only do customers benefit by being thoroughly informed, but employees and managers are part of the company's humanizing marketing approach.

Realizing that the majority of employees are not motivated, despite profit sharing (because if they were, they probably would become tomorrow's competitors!), Ruddick inundates her hundreds of employees with information on the nature and benefits of the products. "Other cosmetic companies train for sales," confirms Mrs. Ruddick, "we train for knowledge."

There are dramatic newsletters that read like underground newspapers, as well as brochures, videos, posters, training programs, and seminars. Delivery trucks are emblazoned with the message: *If You Think Education is Expensive, Try Ignorance*. Of course, everything is printed on recycled paper.

Anita Ruddick is an exciting personality and she transmits that drama and excitement to her staff. When she took on the Amazon Rain Forest as a cause celebre, she had 250 employees ferried to the doors of the Brazilian Embassy in London to launch a protest against the destruction of this global resource. Naturally, she made sure the press were there with all equipment on "Go!"

Social action is part of *The Body Shop's* carefully orchestrated business strategy. Anita Ruddick is frank to admit that her dramatic campaigns produce sales and profits. When she located a soap manufacturing plant in a poverty-stricken section of Glasgow, Scotland, it was a well-publicized event. Poor, largely unemployed people were employed, trained, and turned into loyal consumers. Corporate activism on behalf of the environment has been developed into profitable entrepreneurship by *The Body Shop*.

The Body Shop is perhaps one of the most dramatic examples of eco-entrepreneurism on earth. And the globe, as well as the Ruddicks, are the beneficiaries of it. "It's paying attention to the aesthetics of business," says Anita Ruddick. Amen.

Capturing natural beauty

There is a rather folksy, family run company in Kennebunk, Maine, that makes a line of cosmetic products. The buzzword here is *natural*. The company is Tom's of Maine, started in the late 1960s by Tom and Kate Chappell.

Among the natural products they produce up there and sell all over the United States are toothpaste (the No. 1 seller), toothbrushes with natural bristles, flossing ribbon, alcohol-free mouthwash, deodorants, and shampoos and shaving creams in tubes. Along the way, Tom's of Maine has developed a "chain" of mail order houses that sell their natural products.

Colgate, Crest, and Pepsodent need not worry too much; still, Tom's is racking up $6.6 million a year in sales, based on the tenet that "natural is better." The primary market for Tom's products in New England, New York, Washington (DC), Los Angeles, and San Francisco (and more to come) is the under-50, health conscious, well-educated consumer. In some parts of their distribution area, Tom's have captured 3.5 percent of the market share.

Meanwhile, Tom's of Maine has marked 20 years in business and its 35 or so employees and two happy owners have proven that small entrepreneurs can grow and profit in an environmentally friendly manner. And not only in America

Global releaf

Aveda is a small company that manufactures hair sprays and other beauty aids that are environmentally responsible, are not tested on animals, and avoid alcohol and hazardous spray cans. To promote their products, they teamed up with the American Forestry Association and are participating in that group's Global ReLeaf program. Customers who buy $10 or more of Aveda products in any of the beauty salons that handle Aveda products are assured that a tree will be planted by the AFA in their name. "Plant a Tree So We Can All Breathe Easier" is the slogan of Aveda's environmentally friendly promotion.

Biodegradable: buried in profits

Much has been written recently about biodegradable products. In fact, even the term *biodegradable* has been shrouded in controversy because most of our garbage today is thrown in landfills and buried, and very little, if anything, actually biodegrades. No matter, biodegradable products are sure to be in great demand over the next decade, particularly for new materials that can biodegrade underground.

Biodegradable diapers make a healthy bottom line

Convenience goods save time for busy people but they are not always environmentally friendly. The year that disposable diapers were invented and started selling by the billions, was a dark day for landfills. While many parents thought that these disposables were tops for their babies' bottoms, ecologists discovered that throwaway diapers were the scourge of landfills.

Although the product might be worn for only a few hours, it will not biodegrade in a deep, dark landfill for many centuries. Cellulose, the material currently used in diapers, requires sunlight and air to break down the chemical constituents. But there is light at the end of the environmental tunnel—*Ultrasponge. Ultrasponge* is a development of a company named IGI Inc. It is totally biodegradable, containing a moisture-activated enzyme that starts the decomposition process within 48 hours.

Rather than manufacturing its own diapers, IGI is offering the technology to major diaper manufacturers, who produce more than $3.5 billion worth of diapers a year. It is a product worth keeping an eye on. Already, the stock of this modest-size, 175-employee company from Vineland, New Jersey has reflected this innovation. In the coming year, IGI's bottom line will no doubt rise to the top.

Medical field: biodegradables and BOPs

It is unlikely that an entrepreneur would decide one day that he or she would become a doctor or dentist. More likely, a chemist, a biologist, or a veterinarian becomes the developer or producer of new healing devices.

The field of biodegradable devices and materials has made quantum forward leaps during the past three years. BOP, or biocompatible osteoconductive polymer, is one such product—a glue that can take the place of pins and staples in joint attachment procedures or bone fractures. This new material, a copolymer, was actually developed in Russia. One of its primary uses is in hip replacement operations, shoulder stabilizations, and ligament reattachments. It stays cool as it sets and expands, resulting in a tighter fit.

The BOP polymer, and other materials being developed, biodegrade in the body. It eliminates repeat operations when fractures fail to heal and future operations to remove pins, staples, or bone grafts. BOP's are like scaffolding that surrounds a bone or joint. Not only do biodegradable polymers eliminate later operations, grafts, and possible infections from hard metals, but may be used in the treatment of bursitis in replacement joints in the near future.

Companies like Johnson & Johnson Orthopedics and Ortho-Tex are working on these innovations. The stocks of many other as-yet small biotechnological firms are growing healthily. It is a highly specialized but potentially lucrative area of twenty-first century entrepreneurship.

Coffee brews savings

Most coffee that is brewed is filtered through snow-white paper filters. They cost only about $10 per year in the average household but they are costly to the environment. The filters are bleached white by chlorine, one of the most serious pollutants in our ecology.

Using, and promoting the use of natural, chlorine-free, oxygen-processed filters or permanent mechanical filters, even reusable cloth filters, can make your coffee taste even better and make an environmental contribution.

In Waterbury, Vermont the Green Mountain Coffee Company sells all types of exotic coffees as well as unchlorinated filters. Green Mountain sells a lot of them, especially so since they run a number of busy coffee shops and are franchising others all over New England and elsewhere. The irony is that environment-conscious Dan Cox, one of the founders of Green Mountain Coffee, spent months trying to buy these ecologically attuned filters. He finally found them in Canada.

Another possible by-product from coffee is coffee grounds. Grandmothers used to say that old grounds make plants new. Perhaps the caffeine left over in the grounds stimulates plants as it does humans; perhaps it is just an old grandmothers' tale. But then—it could be an entrepreneurial opportunity of huge proportions and profit.

Conservation store: environment on sale

Ed Lowans Jr., a Toronto architect-builder is heavily into the environment. He builds or rebuilds houses for people with allergies and chemical sensitivities. His latest innovation is a store that sells both retail and by mail order everything that could possibly be environmentally friendly.

In addition to the expected health foods and natural drug products, The Conservation Store in Toronto handles energy conservation products, recycled paper goods, computer light screens, and many more items.

Wholesalers are also accommodated. Suppliers who have the type of products Lowans feels are beneficial to the ecology and his type of customers are wanted. Past experience has taught Lowans that as long as the cost of energy remains fairly high, operations like The Conservation Store will do very well and will no doubt increase and proliferate.

Consulting: getting greener all the time

As people become more ecologically aware, there is apt to be confusion over what course to take to correct environmental injustice. Enter the eco-consultant. In almost every aspect of American life—business, entertainment, home, etc.—lurks opportunity for the smart green entrepreneur.

Electronic Data Interchange

Office waste paper is one of the nation's largest components in landfills, and perhaps no single company contributes more to the mess than the U.S. government. It was a natural step, therefore, that some efficiency and environmentally conscious person would come up with a countermeasure. That person was Congressman Esteban E. Torres from California. Also aiding in the investigation of solutions to the paper avalanche was Congressman Norman Sisisky of Virginia.

During the summer of 1990, the Small Business Subcommittee on Export, Tax Policy, and Special Problems held hearings on the topic "Electronic Data Interchange: Key to Small Business Competitiveness (EDI)." What came out of the hearings and deliberations were recommendations that could bring "paperless offices" to American businesses by the year 2000.

One small North Virginia contractor went the EDI route last year by acquiring a personal computer, printer, modem, and appropriate software. The system was applied to purchase orders, change orders, confirmations, and electronic invoicing. Within the year, the owner noted, his company was able to expand much faster, reduce paperwork and filing, and all at a cost of less than $4,000.

Government representatives who are pushing for EDI feel that 75 percent of government business can be handled electronically by the year 2000, saving mountains of paper. Rep. Sisisky stated, "Small business must join the telecommunications revolution in order to maintain a competitive advantage . . . We want to make certain small business can fully participate in what may be the most important tool for business expansion since the telephone."

Teaching small businesses how to become computer-efficient is an opportunity for a consultant with the right know-how.

Home heating consultant: saving energy, making money

Intelligent, concerted efforts can be used by a home energy expert to save money for home owners and even entire municipalities. Fees could be based on a small percentage of actual savings and commissions on new furnace, window, and insulation sales. The following 10 items could make a productive and profitable, energy-saving checklist—at virtually no cost or even for free, to the home owner:

1. Get a free energy audit from the local gas and electric company and learn exactly how and what they do. For more sophisticated "home auditing," a thermogram (infrared photograph) of the home can be taken. This will pinpoint the exact areas of heat loss, but will, of course, add to the cost of the audit.
2. Check the thermostat and set it back to 68 or 70 degrees during the winter months. Wearing a light sweater is more economical than turning the heat up for the whole house! During the hot weather months, keep the thermostat at 76 to 78 degrees. Wearing fewer clothes while inside also is more economical than turning down the air conditioning!

3. Watch door openings and closings. There might be many occasions during the day when doors are kept ajar and heat (or air conditioning) escapes. If there are kids in the house, get them trained to understand the reason for keeping doors closed.
4. Unused rooms should be sealed off, especially along the bottoms of doors. Keep vents in these rooms closed or shut off the heat, if feasible.
5. Hot bathwater is a useful "heater." A tub full of hot water (about 95 to 100 degrees) can generate about 10,000 BTU of heat while it cools down to 70 degrees. Leaving the bathroom door open after a bath will disperse this energy into the rest of the house. *Then* the drain plug may be pulled.
6. Kitchens generate the most heat in the house. The stove, refrigerator, freezer, and range produce heat. If a door exists that blocks this heat from spreading into the rest of the structure, open it and let the warmth out. Conversely, during hot weather, keep the doors closed.
7. Laundries also generate heat. During cold weather, keep the door to the laundry area open; close it during hot weather.
8. If facilities exist to air-dry some clothes, hanging them out to dry will save considerable dryer energy, as well as distribute humidity throughout the house when winter heating tends to make the air too dry.
9. Showers consume a lot of hot water—35 gallons for the usual five-minute shower. Replacing the old, inefficient shower head with a more efficient, low-flow one can save as much as 50 percent of the energy.
10. Toilets are other water abusers, using five to seven gallons for each flush. A one-gallon displacement bottle in the tank can save on an average eight gallons a day—nearly 3,000 gallons a year. An ultra low-flush toilet (ULF) utilizes compressed air and a mere gallon of water per flush.

In addition to these items, there are innumerable other energy-savings areas that a thorough Home Heating Expert can discover, especially in insulation of windows, doors, wall outlets, vents, attics, etc. Periodic maintenance of appliances and heating and cooling equipment also reduces energy consumption, but the greatest savings can be achieved by educating the home owner and family. Did you know that . . .

- If the American home owner could increase the energy efficiency of his home by an average of 20 percent, we could reduce demand for electricity by the output of 25 power plants.

- If nobody would turn on their washing machine until they had a full load, an average of 45 gallons would be saved with each cycle—billions of gallons nationwide.
- If your refrigerator is set for 32 degrees (when it should be 38 to 42 degrees) and the freezer for −5 degrees (instead of 0 to 5 degrees), energy consumption can increase as much as 25 percent.
- If the gas stove or oven has an electronic ignition system instead of a constantly burning pilot light, about 40 percent of gas can be saved.
- If every homeowner would raise air conditioning settings by 6 degrees (lets say from 72 to 78 degrees) energy used by generating plants to make electricity—nearly 200,000 barrels of oil—would be conserved each day—18 million barrels during a three-month summer season.
- If you replace a 75-watt incandescent bulb with an 18-watt fluorescent one (giving the same amount of light), the latter lasts 10 times as long, saves many dollars on the electric bill, and helps keep 250 pounds of carbon dioxide out of the air (needed to produce the additional energy a regular bulb requires).
- If an old, inefficient oil-fired furnace is replaced with a modern, efficient one, giving 87 percent efficiency (as high as present-day models will go), about $700 is saved during the course of a northern, six-month cool-weather season. The new furnace costs about $2,000 and is thus paid for within three years—the rest of the years being money in the home owner's pocket.

Energy Publications

The Most Energy-Efficient Appliances and *Saving Energy and Money With Home Appliances* ($3 each), The American Council for an Energy-Efficient Economy, 1001 Connecticut Avenue, NW, Washington, DC 20036.

Heating Systems, Public Information Office, Massachusetts Audubon Society, Lincoln, MA 01173.

"Consumer Guide to Energy-Saving Lights," ($2) from *Home Energy Magazine*, 2124 Kittredge St., Berkeley, CA 94704.

Tips for Energy Savers, Department of Energy, Conservation and Renewable Energy Inquiry & Referral Service, P.O. Box 8900, Silver Spring, MD 20907, (1-800-523-2929).

"Cool Aid" Report, *Consumer Reports Magazine,* July 1989, pages 431-440, later editions.

Cut Your Electric Bills in Half, by Ralph J. Jerbert, PhD; Rodale Press, 33 E. Minor Street, Emmaus, PA 18098.

- If an old, gas-fired furnace that is only 50 to 60 percent efficient is replaced with a more efficient, new warm-air furnace giving as much as 97 percent efficiency, delivering about 80,000 BTU/hr., 50 percent of the gas consumption can be conserved.
- If a leaky faucet drips but one drop every second, it can waste 200 gallons a month as well as a lot of heat-energy if it's a hot-water faucet.
- If all small towns in America would adopt a general energy audit policy, they could save millions of dollars each for needed local improvements. A town of 3,600 inhabitants, Osage, Iowa did just that—replaced inefficient furnaces, insulated leaky windows, wrapped hot-water heaters in insulating blankets—and saved $1.2 million in energy costs during a single year.

Consulting industrial waste disposal

In 1984, a young promotion and marketing expert named Frank Stefano saw the handwriting on the wall: the landfills were filling up; solid wastes were piling up; and the people were fed up with dangerous dumps. Stefano swung into action and formed Environmental Waste Technology (EWT) in Newton Upper Falls, Massachusetts. He has since been consulting with private, commercial, and municipal waste companies and agencies.

EWT now has hundreds of projects annually throughout New England. EWT's job: to find the best ways to dispose of industrial wastes and hazardous products. From the looks of it, the company is not likely to run out of projects, problems, or customers very soon. It is another example of how environmental needs and entrepreneurial spirit can sleep together in a very profitable bed.

Office management: managing the environmental way

If you are an office manager, you are right on the environmental firing line. If you want to be an environmental consultant, office management is a good place to start. Matthew Costello founded Corporate Conservation and did just that. J. Rodney Edwards of the American Institute advised giants AT&T, the Bank of America, and 3M on office paper recycling. These companies were able to save $485,450, $354,000, and $1.5 million respectively on the sale of their office waste paper alone, not even counting on how much more they saved on unneeded waste pick-up costs.

Each ton of waste paper recovered for recycling conserves 3.3 cubic yards of landfill space. Waste paper makes up 40 percent of landfill space. Each office worker averages 180 pounds of high-grade recyclable paper each and every year. According to the Environmental Protection Agency, an office that recycles its office paper can lower its trash volume, and cost, by an average of 34 percent. A ton of paper recycled

Do-It-Yourself Office Recycling

1. Evaluate the recyclable material that can be generated in the office, including paper (mostly white), plastic, and aluminum. Contact the EPA for a manual on office evaluation (see nearest office in chapter 9. Eighty-five percent of office waste is recyclable paper.
2. Locate a market for your recyclable waste, either in the telephone book or if you are in a large building, through your in-house custodian.
3. Plan an employee educational campaign. Appoint coordinators and get periodic feedback on results.
4. See whether, in many cases, both sides of a sheet can be used, or whether the back of surplus and outdated forms can serve as scrap paper or file paper.
5. Is it possible to organize car pools to conserve auto and gas use?
6. Check, or have checked, the air quality and try to make the office atmosphere healthier. This includes eliminating smoking within the office.
7. Look for water stains on the ceiling or wet spots in carpeting that indicate a water leak somewhere—and possible bacterial growth.
8. Are all bulbs fluorescent? They give off less heat and use less energy than incandescent bulbs.
9. Set out clearly marked containers to collect other-than-paper waste products: one for glass, one for aluminum, one for newspapers. (It is interesting to note that we throw away enough aluminum each year to build four complete new airfleets or that if we stacked up all the writing and office paper used in the nation's offices, it would form a "Great Wall of China," 12 feet high, reaching from Los Angeles to New York.)
10. Since virtually every office and shop makes coffee or tea, buy supplies in bulk. You save that 10 percent that normally goes for packaging and cut down on food-wrap waste, which makes up as much as one-third of all our landfill waste.

And for good measure, if every employee has his or her own cup or mug (what a wonderful Christmas gift idea!), you would not need any styrofoam cups, which cost money, crowd up the wastebaskets, and are nonbiodegradable.

saves 24,000 gallons of water and 3688 pounds of wood (about 17 southern pine trees), which are needed to manufacture new paper.

Coca-Cola USA of Atlanta realized these statistics and began a recycling program in its corporate offices. The result has been more than 1.5 million tons of solid waste diverted from landfills. The recycling effort has won praise from Keep America Beautiful and the American Paper Institute. With top management's support, the program is accelerating each year, utilizing bulletin boards, newsletters, posters and, of course, new recycled paper stationery.

Employment agencies: focus on the environment

Perhaps it has yet to happen, but there is a notable surge of jobs available in environmental categories—both in organizations that work directly on environmental concerns and within corporations that deal with the environment or are becoming increasingly conscious of the environment.

Engineering firms, for instance, are looking for environmental engineers, hydrologists with an environmental background, wastewater engineers, and geotechnical managers with environmental experience. On a typical Sunday, the *Washington Post* carried large ads for one of these positions by Baker Engineers of Beaver, Pennsylvania and Atlantic Research Corporation of Gainesville, Virginia. In a related category, Labat-Anderson Inc. of Arlington, Virginia was searching for environmental scientists, aquatic ecologists, environmental policy specialists, biotechnologists, environmental modelers, agricultural chemists, and toxicologists.

Sixteen other ads asked for a variety of other environmental specialist—including consultants, research analysts, service managers, and sales representatives. Even the Environmental Protection Agency had placed an ad for several positions.

Perhaps here is an entrepreneurial opportunity for a well-versed and connected environmentalist to open an employment placement agency, specializing in the g-r-o-w-i-n-g field of ecology and environmental protection.

Environmental audits

Instead of buried treasure, some bank-supported building projects have dug up hitherto unknown toxic dumps. Such unwelcome serendipities have cost banks millions of dollars in mandated cleanup operations. Even picking up a defunct savings and loan could be hazardous to a bank's health if one or several of the mortgaged properties turns up to be polluted and in need of detoxification.

Senator Jake Garn of Utah took one look at this mess and said, "There's got to be some sanity out there." So, he promptly introduced a law that would limit the Environmental Protection Agency's ability to sue lending institutions to recover the costs of toxic cleanups. The senator proposed that the Environmental Protection Agency be granted more money for needed cleanup efforts, but that Savings & Loan crisis is complex enough without adding yet another dimension to their insolvency and the bill languished.

In hundreds of cases, however, the cost of insolvent S&Ls and banks is less than the cost of an Environmental Protection Agency-mandated cleanup of toxic or hazardous conditions. Today, in virtually all cases of mortgage lending, an environmental audit of the property should be undertaken. This opens up a whole new entrepreneurial category.

In today's litigious climate, corporate acquisitions and mergers can create unseen, potential disasters for unsuspecting buyers. Hazardous, and as yet, unknown, underground wastes, groundwater contaminates, radioactive construction materials, defective underground storage tanks, etc., are all real concerns. This is good news for lawyers specializing in environmental law, however, and who must be familiar with such environmental ramifications and even be able to estimate potential insurance costs to protect buyers against conceivable claims.

It is difficult to obtain accurate estimates of pollution cleanups, however. Contractors are loath to give firm figures without digging into the ground or having a thorough soil analysis conducted. Consequently, even tentative tests can cost many thousands of dollars, paving the way for entrepreneurs who can offer reasonable environmental audits.

The number of environmentally sensitive deals will continue to rise in the coming decade, providing opportunities for those entrepreneurs who are quick to recognize the high cost of cleanup and who can capitalize on it in ingenious ways.

Entertainment: eco-style

The Wetlands Preserve is an "eco-bar" that started in New York City early in 1989. Festooned in tie-dyed cloth and beads, it has been attracting large crowds and more than a dozen special events a month. "The intention of the club," says general manager Brian Gibson, "is to put together a place where people interested in environmental issues can blend in a social environment."

During Halloween, the club featured a "Trick or Trees" night, giving out seeds to all so that they might keep New York green. Music is typically heavy rock groups, including some famous ones like The Grateful Dead and Sting, who are identified with environmental causes.

The club even has an "eco information center"—a parked VW van that serves to dispense information on the environment. A special eco-organizer plans the events for this swinging eco-saloon.

It's an idea whose time has come and that can be "planted" almost anywhere in America. What a great place to recycle aluminum beer cans. Who'll open the first Greenpeace Pub?

Television: the "greening" of the tube

Opportunities in television production and writing are popping up like crocuses in spring thanks to Hollywood producer and environmental activist Norman Lear, who, in 1989, was able to recruit the entertainment division heads of ABC, CBS, and NBC to sit on his board of the Environmental Media Associates.

The EMA serves as a clearinghouse for ecological information. The Association's contacts have enabled it to work environmental messages into numerous television movies and more than two dozen prime-time series, and it is just the start.

Among the outright environmental shows and others that include ecological information are:

- *E.A.R.T.H. Force* on CBS is sort of an environmental A-Team adventure that contains some dubious antibusiness footage.
- *America the Beautiful* on ABC, a conversation between President Bush and Curt Gowdy about the United States's natural wonders.
- Bill Moyers' *Global Dumping Ground* and Frontline's *Decade of Destruction* are aired on PBS.
- *Race to Save the Planet* is another PBS presentation in 10 parts, starring Meryl Streep and Roy Scheider.
- *Ground Zero* appears weekly on VH-1 and zeroes in on rock celebrities like Sting.
- *Voice of the Planet*, a New Age miniseries, *Network Earth*, a news magazine format show, and *Captain Planet and the Planeteers*, an animated cartoon series of superheroes and Mother Earth that is aimed at teenagers—all on the Ted Turner network (TBS cable).

The danger in any sudden neo-movement is that the pendulum often swings too far in a radical direction. Some ideas are unbalanced, fantasies culled out of ardor rather than reality, and often counterproductive.

Consequently, the opportunity for environmentalist-writers and producers to create veritable, logical, scientifically proven scripts is a real one indeed. By working together with others concerned about the environment, both the heroes and the villains, a new Voice of the Planet can be generated. Remember the story of the one rotten apple in a bunch? That's why *all* the apples, all the people on earth, must be convinced to work together. Leave adversary programs to the lawyers; let environmental programming promote togetherness. A transcript of the documentary *Profit the Earth*, a popular television documentary on the marriage of environmentalism and entrepreneurship, appears in Appendix E.

Fashion farsightedness

There's been enough publicity about killing wild cats for furs, elephants for ivory, and baby goats for wallets to give a person a real guilt complex. Fashions and fashion accessories can now be manufactured—and sold—without any guilt whatsoever, however. Today, these items are being made from substitute materials.

Carolee Designs of Greenwich, Connecticut makes resin jewelry that looks remarkably like real ivory. They promote their line of necklaces, bracelets, and earrings by giving "save the elephant" parties for store buyers and donating 10 percent of their sales to elephant preservation organizations.

Lena Fiore Inc. of Cleveland, Ohio sells "fur coats" that are made of, believe it or not, leftover turkey feathers. With all the turkeys consumed in the United States, feathers are no problem. Lena Fiore buys up tons of feathers, ships them to Italy where they are processed into furlike fabric and high-fashion coats, then imports them back to the United States where fashion-conscious ladies are said to be gobbling them up.

In Juneau, Alaska, Alaskins Leather does the Southeast Asian eel skin business one better. They have found a way of processing salmon skins, purchased as otherwise surplus material from canneries, and making them into wallets, belts, and other small leather goods. Alaskins find nothing fishy in this conversion—it's environmentally beneficial as well as entrepreneurially profitable.

A Vancouver, British Columbia company operated by Stephanie Seaton, has been creating attention-getting fashion shirts for the past couple of years. Producing active sportswear with a difference has given the young entrepreneur a $350,000-a-year business.

What sets her designs apart from the run-of-the-mill fashions are two factors: (1) all designs depict colorful and attractive drawings of wildlife; and (2) 10 percent of the company's profits are divided between the Western Canada Wilderness Committee and the Amazon Rain Forest funds. Miss Seaton's company, Hycroft Collection, Inc. has received additional leverage by merchandising its designs through sports clubs.

Clothing recycling

Well beyond the pedestrian racks of Good Will stores are high-class fashion emporias that purvey yesterday's high styles or outgrown kiddie clothes, or simply modes of which Mrs. Astor has tired. On New York's upper Madison Avenue, fashion stores have waxed profitable for many years. In most metropolitan suburbs across the country, second-time-around boutiques are flourishing in smaller shopping centers or nooks and crannies in low-rent locations, sometimes run by an enterprising businesswoman, sometimes by a nonprofit organization.

In the last few years, more and more children's clothing and toy stores have made their appearance. Resale of men's, women's, and children's designer-label fashions has become unabashedly big business. In and around New York, at least a dozen top-fashion kiddie shops can be found. The selection of merchandise is usually top-notch and not always cheap—though well below the original price tags. All merchandise, before it is accepted into these often red-carpeted boutiques, has to be cleaned and pressed and in near-perfect condition.

Sometimes merchandise is accepted on consignment—that is, it is held for resale for a period of time, generally 30 to 90 days, and, if not sold, returned to the owner or given away to a lower-level charity—sometimes purchased outright at a fraction of original costs.

Each resale or recycling store determines its own policy regarding consignment or outright purchasing, length of time merchandise is held before it is sold, on-premise repairs made, pass on to a charity or return to seller, pricing and markup, etc.

It is a good business, however, that enables those who can afford better clothes to clean their closets, for buyers to obtain better merchandise at lower cost, and for entrepreneurs to go into a business that often involves very little up-front capital. There is one other aspect to remember: used clothing can become a valuable export item to lesser-developed countries.

Forestry: seeing the green because of the trees

Forestry has long been a hotbed between environmentalists and developers, timberman, and other industries. Environmentalists value trees for their nontangible contributions like conserving water, curbing runoff, and purifying the air. Smart entrepreneurs, however, can find value in conserving, replenishing, and replacing one of our most valued resources—our forests.

Lumber: entrepreneurs up a tree

The lumber business is not an easy one to go into, especially the cutting and processing of timber. It is a gigantic industry dominated by companies like Weyerhaeuser, Crown Zellerbach (now owned by Great Britain's Goldsmith interests), Plum Creek Timber Co. (owned by Burlington Northern), Pacific Lumber Co. (owned by Maxxam Corp. of Houston), and about 1,400 other major lumber companies.

Environmentalists and lumbermen have always been at loggerheads. In recent years, as large companies have been threatened or even swallowed up by even larger ones, a strange phenomenon has emerged. The smaller timber companies received support from environmentalists. The reason for this dog-loves-cat symbiosis is that smaller companies generally practice long-term growth, replanting older trees with new ones as the adult ones are thinned out.

At least 70 percent of all timber lands are owned by private companies; 30 percent by governments. It is estimated that 90 percent of timber land owners practice sustained yield—that is, they replenish old trees that are cut down with new ones. After about 10 years, depending on the growth cycle of the specie, the trees are harvested. Groups like Forests Forever and the Native Forest Council are trying to stem rapid cutting, especially in that five percent of forests that are still considered virgin, or the original stands, some many centuries old. Virtually the only virgin timber still standing is on federal land.

Loggers, particularly large ones governed by balance sheets and investors rather than by environmental concerns, consider the forest in more pragmatic terms. "A young stand of trees is much better than old rotten trees," said a chief forester with Weyerhaeuser Co. He reflects the timber industry's position that the old forests are a waste of wood, because timber allowed to die off and rot is neither economical nor beautiful.

Environmental groups led by the American Forestry Council are following the successful lead of Israel by advocating the private and institutional planting of millions of new trees. Israel has planted more than 150 million trees, primarily through a quasi-government organization called the Jewish National Fund. Its purpose is to replenish the badly denuded soil, hold it for restoration, bring back wildlife, and at the same time, develop an industry for its burgeoning population—furniture, construction material, energy needs, as well as aesthetics and carbon dioxide absorption.

Some entrepreneurs have either bought or planted small timber stands. Some are for replantable seedlings, some for Christmas trees. It is not a business for short-term profits but for long-term development. While it is true that "only God can make a tree," reforestation is also a big business—indeed, a growing business.

Funerals and forestry: living memorials

Funerals and forestry seem like a unique way to help the environment, but the Batesville Casket Company of Batesville, Indiana has been doing just that for more than 15 years. With every Batesville casket purchased, the company arranges to have a tree planted as a living memorial to the departed.

The name of the environmental effort is the Batesville Living Memorial Program. The U.S. Forest Service is the federal agency that guides the program and implements it, and Batesville picks up the cost. To-date, more than 5 million trees have been planted throughout the United States. In Cincinnati, Ohio an entire blighted area was reforested, and throughout 1990, nearly half a million trees were planted.

A forest service official stated that "carbon is known to be a major contributor to warming of the earth's atmosphere . . . each tree planted can remove five to ten pounds of harmful carbon from the atmosphere during each year of active growth."

Every undertaker in this country—and there are nearly 22,000 of them—can offer just such a program to his clients. Millions of living memorials would be a wonderful, environmentally beneficial heritage to leave behind.

Woodworking

The Hummers of Texas are a harmonious family enterprise. Idealism and a sense of shrewd Yankee tradesmanship blended with real concern for the environment and their customers—that's the unique combination of the Hummers. Located on a spread deep in the heart of Texas, where mesquite grows wild and wonderful, this family has developed a diversified $100,000 business with loyal customers in every state and all over the world.

Half of their products are homemade from wood products that have fallen down, not cut; the other half are homemade by other craftsfolk with the same penchant for serving the environment, pleasing themselves, and eschewing the cynical caveat emptor of commerce.

The Hummers have been creating a whole passel of precious wood products since 1976, working 12 hours a day. While the exquisite world of wood artistry is the mainstay—vases, bowls, walking sticks, benches, and little chests with secret drawers—The Hummers also produce small inspirational books to satisfy father's literary penchant, and all types of closet and stocking stuffers made of cedar and juniper. One notable aspect of these enterprising folks' business is this: meticulous attention to recordkeeping, customer satisfaction, and scrupulous environmental integrity. They won't ever get rich, but they'll live happily, thanks to the wonder of wood. The Hummers, Inc., Reagan Wells Canyon, Box 122, Uvalde, Texas 78801.

Most soil needs chemicals and bacterial additives. Agronics Incorporated of Albuquerque, New Mexico saw the need for a series of 100 percent organic or natural soil conditioners and has been producing them since 1967. Today, Agronics does more than $1 million dollars worth of business and has expanded beyond the U.S. boundaries and to as far away as England and Taiwan.

Gardening supplies: green thumbs beget green

There is money in them thar frills, especially the floral ones that are keeping millions of home owners busy and more than 15,000 garden supply stores in business. Add to these numbers, more than 800,000 people who have bought flower and vegetable seeds from Burpee, and more than a million people who subscribe to organic gardening publications, and you have a formidable number of green thumbers who are keeping entrepreneurs in this business "green."

Several very large and successful garden supplies companies that focus on environmentally friendly merchandise, top-quality tools often guaranteed for a lifetime, and heavy national promotions have become known. But there are also many small ones that do business around a million or as little as $100,000 a year.

Magic Garden Produce in Woodstock, Virginia is one of those little garden suppliers that grows organic watercress, tomatoes, and other produce but also raises range-fed chickens and pond-fed trout. They ship—primarily wholesale—to dealers in five states and make a happy, ecologically friendly living on their own local areas.

Another small operator is Wooden Shoe Gardens of Cincinnati, Ohio, which sells organically grown products to mail order customers in three nearby states. Produce, garden supplies, and groceries. Since 1976, a loyal cadre of followers have been buying from Wooden Shoe.

Hazardous waste disposal: profiting twice

The tragic Valdez oil spill at Prince William Sound in Alaska is dwarfed 20 times over by our nation's nicest guys—our neighbors. The millions of do-it-yourselfers who change their own motor oil dispose an estimated 200 million gallons of old oil each year. According to the Environmental Protection Agency, just one single quart of the icky black stuff, dumped down a corner sewer while nobody is looking, can seep into a municipal water supply system and make 250,000 gallons of drinking water taste like you-know-what. Enter Donald R. Smith, who owns four quick-lube drive-ins in Rhode Island and Massachusetts, and another Valvoline franchise owner in Lexington, Kentucky, who say: Give us your old, your black, your tired oil, and we will help you save the earth.

Indeed they do, and so do many other professional auto services. There are a number of things that these entrepreneurs can do with the old oil: (1) have a trucking company haul the old oil away periodically for proper disposal; (2) sell it to local greenhouses who can use it to heat their facilities during the cold-weather season; or (3) install heaters on their own premises that can burn waste oil and keep garages warm.

Each oil change operation disposes between 1,000 and 3,000 gallons of old oil every month—12,000 to 36,000 gallons a year. Valvoline alone, in its nearly 300 franchised centers, is helping to recycle an estimated 6 million gallons of old oil each year.

Nationwide, it is estimated that only 10 percent of old motor oil is being recycled. Alabama, however, is doing something about this. Project ROSE (Recycled Oil Saves Energy) was started at the University of Alabama in 1977 by a chemical engineering professor. Today, more than 35 percent of all 6 million gallons are recycled—35 percent more than the national average. The entire state, its schools, public utilities, service stations, and municipal garbage collectors are cooperating in keeping polluting oil out of the Alabama earth and water system. In some cities, even the refuse pick-up trucks carry containers in which do-it-yourselfers and garage owners can dump old oil.

Any entrepreneur interested in the used motor oil recycling business can get more information from these two sources:

U.S. Environmental Protection Agency, OS-323
401 M Street, SW
Washington, DC 20460

Project ROSE
P.O. Box 6373
Tuscaloosa, AL 35487-6373

Bioremediation gobbles up garbage

Bioremediation is a mouthful. Actually, it is millions of mouthfuls. The term is a process called microbial cleanup and it is hardly a new idea. Sewage treatment plants have been using bacteria to filter waste from water for decades. Only recently have the concerns for the environment and economics searched for a commercially feasible method to clean up real or threatened hazardous waste problems. Using proven bacterial intervention proved to be the answer.

The United States generates 600 million tons of hazardous waste and wastewater each year. For the year 1992, it is estimated that the cost of disposing and treating these waste products will climb to $80 billion. The size of the problem, or *market*, as it were, has attracted more than 100 companies that are battling this growing pollution problem with the help of a thousand different species of fungi and bacteria. The enemy is gasoline, PCB, vinyl chloride, pentachlorophenol, and a host of other chemicals that are the by-products of our civilization.

The antidotes, the bacteria, work cheap. All they need is the natural nutrients found in soil, oxygen supply, and lots of toxic substances. By a process of natural selection, the microbes replenish themselves quickly and only the hardiest ones survive to tackle more toxicity. In the case of pentachlorophenol, the chemical used by wood treatment plants, a super-microbe called *Flavobacterium* was developed in a University of Idaho laboratory.

Dealing with bacteria is by no means simple or a job for amateurs. While most bacteria require oxygen to function, others such as anaerobic bacteria, require oxygen-restricted environments. Many problems are solved in laboratories, but in real-life situations, they don't always work out.

Clearly, not every cleanup job is a candidate for bioremediation. For instance, bacteria will not work in dense soil made up of clay. Cleaning up PCB hazards also are extremely difficult, because no microbes have been found that can attack the molecular structure of this chemical, a known cancer-causing combination.

Where bioremediation can work, it is a godsend. Contaminated soil can be cleaned up for $30 to $50 per cubic yard. Incinerating or hauling away the same amount can cost $200 or more. The Environmental Protection Agency has identified 1,219 sites, just on their priority list, that require cleanup, and the number of sites is expected to double by the end of this decade.

Not all sites will be cleaned by microbes, but enough of them will be handled by the bioremediation companies to make this another environmental growth industry.

Co-processing: cementing success

Of the 265 million tons of hazardous waste generated in the United

States and Canada each year, much of it winds up being burned or incinerated. Many people are against this process because the smoke and ashes generated in the process escape into the air and land who-knows-where. Of course, there are technologies that can capture the hazardous residues, but there is also a better way—co-generation.

Co-generation, such as is practiced by the seven Systech Environmental Corporation plants*, the largest operation of its kind in the United States, is a co-processing technique. Co-processing is using the cement manufacturing process to recycle, reuse, or treat waste while simultaneously manufacturing a product in a single, combined operation.

Co-processing is big business, and it is a cost-effective alternative to landfills. While landfills bury our wastes and thus presumably get the garbage out of sight, we are not only running out of space, but the very wastes in the earth can prove troublesome to underground water supplies and future building projects in the area.

High-temperature incineration (3000 °F or more) used in this cement-making process virtually eliminates toxic side effects, use of valuable land, and even produces a useful by-product—energy.

Co-processing in cement kilns has some limitations, however. Not all types of solid wastes can be burned. To maintain the immediate and potential benefits of co-processing, selective acceptance of waste products must be enforced. The waste material that is processed winds up in cement and cement block. Obviously, any component of this building material cannot contain ingredients that could prove hazardous at some future point in time. Cement kilns that use waste materials as fuels do not generate any greater emissions than those run on more conventional fossil fuels.

Cement manufacturing is one of the largest mineral product industries in the United States. It generates an estimated annual production capacity of nearly 90 million tons. We produce less than two-thirds of the cement needed in the United States, importing the rest from Canada, Europe, and Asia.

Using waste materials in the cement-making process tends to reduce costs and gives American manufacturers an edge over foreign competitors. Equally important, co-processing substitutes scarcer and costlier fossil fuels with waste-derived fuels, reduces energy costs, and destroys much hazardous waste.

Specifying cement and cement products produced in American co-processing facilities is an environmental contribution.

---

*Demopolis, AL; Lebec, CA; Greencastle, IN; Fredonia, KS; Alpena, MI; Paulding, OH; and Quebec, Canada.

Dry cleaning: a clean era ahead

Revisions of the Clean Air Act name 191 chemicals that are no-no's. Such compounds, many of which are suspected carcinogens, include perchloroethylene, the fluid usually used by dry cleaning establishments.

The problem for most of America's 15,000 dry cleaning plants and nearly 50,000 dry cleaners is that "perc" is the toxic solvent used by the majority of existing dry cleaning equipment. To replace existing machinery with more modern, environmentally approved ones, would mean an average investment of $100,000 per store. It would result in considerably higher dry cleaning prices and perhaps the end of the one-hour service business.

Still, it must be done. The Sierra Club's spokesman, Daniel Weiss, director of environmental quality, said, "Our belief is that the cost of the cleanup is far smaller than the damage done by not cleaning up . . . How do you put a price on a pair of lungs?"

It would be naive to assume that a small corner dry cleaner can put up $100,000 for totally new equipment that will accommodate nonpolluting techniques. The money will have to be raised by borrowing; the payback will be achieved by higher prices. If the machinery can be amortized over five years at about $25,000 a year (including interest), and the average store does $250,000 worth of business a year, this would mean a 10 percent increase in dry cleaning costs.

While entrepreneurs must take the environment into consideration, environmentalists cannot ignore economics. But then this also spells opportunities. If only 50 percent of existing dry cleaning stores replace old machinery with new ones approved under the Clean Air Act, it would mean a $750 million infusion of machinery business. Also, eliminating the solvent "perc" will generate production of a safer chemical or even a water-based method.

Banks have said that they will lend 75 percent of the cost of new equipment, generating much additional banking business. For the entrepreneur, however, it will be a problem to come up with the 25 percent differential. Some small entrepreneurs might not make it.

Of course, entrepreneurial ingenuity has alternatives. One of them is the addition of a charcoal-filtering vapor absorber (costs about $6,000) that eliminates most "perc" fumes. Another is a respirator mask and latex gloves for the operator of the equipment. The Environmental Protection Agency, administrator of the Clean Air Act, has not yet decided whether these measures are adequate, however.

Grease gobbling gremlins

One of the curses of the restaurant business is the accumulation of grease in the drain pipes and grease traps of their kitchens. Short of running a sandwich shop or ice cream parlor, almost every restaurant has this problem. Accumulated grease not only attracts vermin and bacteria, but is a real fire hazard. Cleaning it out is difficult and costly. Until now, that is.

In Campbell, California, a modest-sized company named Bio-Care grows a benign bacteria that feeds on grease. While the restaurants' customers are told that they should abstain from so much fat, these bacteria actually lap up the fatty stuff—for about $250 per month.

Federal and state governments have helped new-age companies like Bio-Care by limiting the use of more traditional grease solvents. Pouring bacteria cultures down drains does the job without deleterious side effects, turning accumulated grease into water and carbon dioxide.

As a modest-size company, Bio-Care has gone the franchise route. The four-year-old California company has sold a couple of dozen franchises for $15,000 a piece. Entrepreneurs who want to let Bio-Care's grease-munching microbes do their work, also need about $20,000 for equipment and proper storage for the bacterial buggers.

Solvent service keeps company very solvent

Safety-Kleen Corporation of Elgin, Illinois is in the reclaiming business in a big way. They started real small, of course, but worked themselves up from 800 customers to 180,000. In the process, they went from grossing $50,000 during their first year to an annual gross of $500 million today.

What Safety-Kleen does is reclaim millions of gallons of cleaning solvents from gas stations, lube shops, auto repair shops, and dry cleaning establishments. They provide each of these places with 15- or 30-gallons barrels with clean, recycled solvents suitable for their needs. While the company is helping to clean up the environment by preventing these hazardous fluids from contaminating the earth, they are also cleaning up financially. Annual profits have been running at least 20 percent a year.

With a solid base in solvents, Safety-Kleen has now branched out into recycling and reclaiming used motor oils. It is not yet a profitable operation, but the market is huge and the prospects are excellent, especially since the price per barrel of new oil has accelerated along with the Middle East's political temperature.

Advertising in lube and service stations that old oil is being recycled so that we can lessen the dependency on foreign oil, as well as help save the environment, should be worth a small add-on cost—let's say 50¢ per $16.50 lube job. It's an entrepreneurial opportunity worth exploring.

Homebuilding: marketing the eco-home

Of course, the first eco-homes had to start in California, especially in Smog Valley, the Greater, Grimier Los Angeles area. Eco-Home Network has erected the Eco-Home Demonstration Home, which offers visitors a look at an ecologically sound lifestyle in an urban environment.

The one-family residence has a front yard landscape that can survive mostly on available rainwater. The sun provides energy for heating and bathing and cleaning water, as well as electricity for the entire home. A food garden and orchard, scaled to city size, is managed organically, using on-site generated compost and very little additional organic fertilizer. The home is open to tours by local businesses and the public. Call (213) 662-5207 for details.

In nearby Burbank, the Eco-Logic Development and Construction Company worked with a documentary production company to produce a very excellent public broadcasting service film entitled *Your Home Our Planet*. The documentary features ways that people can improve their lifestyle and property value while helping to preserve and clean up the planet. The model home includes such ecological amenities as a fish pond, energy-efficient windows, solar panels, and even planters for edible vegetation.

Another home with ecological emphasis is the Ed Lowans Jr. home in Caledon East, Ontario (Canada). It is but one of a number of homes built or remodeled for clients with allergy problems. Emphasis in these homes is placed on environmentally friendly, nonallergenic and nontoxic materials. Lowans, however, is only one of numerous architects and builders throughout North America with similar concerns. These ecologically attuned entrepreneurs got together to form a network, focusing on building or renovating structures for people with allergies and chemical sensitivities. Today, about 100 members subscribe to the environmental philosophy of the Environmental Construction Network.

Home environment specialist

As far as we know, there is no business like home environmental specialist—but there certainly is an opportunity to become a HES. If consultants on closets, fashions, makeup, coloring, and decorating can make a living, why not a person who can advise home owners (and businesses that run offices) on the safety of their environment? We can envision such a business sending trained experts into homes just like exterminators. They will be trained to detect abuses in the use of toxic materials in the home (or office) and make suggestions of safe alternatives. They will be equipped with a checklist of items to look for. It is a fairly easy business to learn and can be franchised at low cost. The following is a pro-

To make one gallon of HES, nontoxic all-purpose cleaner, dissolve in one gallon of hot water 1/4th cup of soap flakes, 1/4th cup of white vinegar, and 1 tablespoon of baking soda. For an extra-strength cleaner, double the amount of ingredients in a gallon of hot water. Use gloves when using this stronger solution.

posed checklist of more than 30 items covering a wide spectrum of typical household products and areas. You can add your own or divide the checklist into areas, such as class 1, class 2, and class 3, etc., and have graduated charges for the various services provided.

1. *Air fresheners.* Easy-to-use commercial deodorizers work by camouflaging offensive odors by deadening your sense of smell. Instead, boil sweet herbs or spices; eliminate the cause of the smell; grow plants; ventilate the area; place a couple of ounces of baking soda in the bottom of trash cans; or place one ounce of baking soda or vinegar in small cups around the house.
2. *Ammonia.* The fumes from this old standby can be irritating to eye and lungs and those with respiratory problems. If mixed with commercial cleaners or bleach, the chemical combination can form deadly fumes. Instead, try borax or non-chlorine bleaches first, or ventilate the room well if you have to use ammonia.
3. *Batteries.* Mercury, lead, and other dangerous chemicals can make batteries dangerous, especially car and boat batteries. Find a battery recycler and return used batteries to him (usually some small payment is given). If you use a lot of small batteries, return these to the dealer for mass disposal as well.
4. *Bleach.* Look for non-chlorine bleach; use about 4 ounces of borax per wash load of laundry as a safer substitute.
5. *Carbon paper.* Traditional carbon paper is safer than the carbonless variety.
6. *Cleansers and cleaners.* To be on the safe side, use old-fashioned cleaners like warm water mixed with soap, fortified with a little vinegar, baking soda, washing soda, or borax. A spray bottle of 1 teaspoon of soap and 2 teaspoons of borax in 1 quart of water makes a good, all-purpose cleaner.
7. *Cockroaches.* These pesky creatures are best discouraged by keeping immaculate environments, disposing of garbage properly and often, and caulking all openings and cracks that might provide entrance. If chemical countermeasures are required, use "roach motels" and nontoxic commercial mixtures usually composed of diatomaceous earth and borax.
8. *Correction fluid.* Warnings on correction fluid bottles tell you that they could be dangerous. Look for those that are water-based (marked "for photocopies").
9. *Deodorants.* Some deodorants are proven irritants to sensitive skin. Borax, or a mixture of this inexpensive white powder,

with a little corn starch or a few drops of coconut oil, or even your favorite perfume, will do the job safely.

10. *Detergents.* Instead of nonbiodegradable chemical detergents, use soap flakes or powder with 2 to 3 ounces of washing soda added per laundry load. An additional 4 ounces of borax will increase the cleansing power of tougher laundry. If you must use detergents, look for "phosphate-free" on the label.
11. *Diapers.* A diaper service using cloth diapers costs no more than nonbiodegradable disposables.
12. *Disinfectants.* Get used to keeping all surfaces clean and dry. For a hospital-style disinfectant, dissolve 2 ounces of borax in a quart of warm water.
13. *Drinking water.* Analyze the source of local water supply. If it is from a private well, have the water analyzed by a laboratory for accurate evaluation. Community water suppliers provide monitoring reports as required under the Federal Safe Drinking Water Act. To check for looks, taste, or pollutants, two checking tools are available (wheels) for $5.50 from Environmental Hazards Management Institute, P.O. Box 932, Durham, NH 03824 and *Consumer Reports Magazine. Consumer Reports* prints analyses on water filters and systems (even simple cartridge bucket filters like *Brita* make water *taste* better and produce better hot beverages).
14. *Drycleaning.* Buy mostly washable garments whenever possible. Items that need dry cleaning can be washed with mild soap and a little vinegar, even if the tag says "dry clean only." If you must use drycleaning fluid, have the room well aired (most such fluids contain toxic perchlorethylene).
15. *Fabric softeners.* In place of costly additives that could be toxic or polluting, use 2 ounces of baking soda or a cup of plain white vinegar in a load of laundry during the final rinse cycle.
16. *Fleas.* Brewer's yeast is a good antidote to combat fleas when added in small amounts to your pet's food or rubbed lightly in its coat. A flea trap can be made as follows: fill a shallow dish with soapy water placed in a flea-infested area; place a bare light above the dish. Fleas are heat-seekers and will jump toward the light bulb, fall into the dish, and drown.
17. *Garden pests.* See *Pesticides*.
18. *Glue.* Hobby glues and rubber cement contain volatile solvents that emit toxic odors. Use stick-type or white glue instead.
19. *Insect repellent.* The best repellents are cleanliness and locating the entry points of unwanted insects. Small pools of accu-

mulated water are breeding grounds for many bugs. For a safe, homemade, inexpensive foliage spray, use a spray bottle filled with one quart of water in which two tablespoons of plain liquid soap have been dissolved. Other nontoxic pesticides include (read the labels!) bacillus thuringiensis, pyrethins without other additives, or diatomaceous earth. Keep in mind that some insects are beneficial. In fact, ladybugs, praying mantises, and even many spiders, are your finest allies.

20. *Laundry.* Heavily soiled items can be presoaked in warm water and about 4 ounces of washing soda for half an hour. Rubbing stubborn dirt spots with liquid soap prior to laundering will also loosen dirt.
21. *Markers.* Most markers contain toxic solvents. Use wax pencils or crayons instead for almost anything that needs to be marked.
22. *Medicines.* Medicine taken upon doctor's orders, can save lives. Once past their recommended life, however, they could turn dangerous. Check expiration dates on bottles in medicine cabinets and dispose of old ones—preferably turning them back to the pharmaceutical dispenser for disposal.
23. *Moths.* The first step in preventing moths is to make sure all stored clothes are clean and dry. Once thoroughly clean, they can be stored in tightly sealed boxes or cedar closets. Rather than toxic, commercial moth repellents, make equally effective sachets to place or hang among clothes. These can be made of cedar shavings, lavender, or a mint/rosemary herb mixture.
24. *Oil* (used, motor). Do-it-yourself oil changers dump more old oil into the ground or sewers than the *Valdez* spilled off the Alaskan coast. (Exxon, please note: you missed a compensatory PR opportunity!) Old oil should be turned over to a gas station that does periodic recycling. Dumped oil can wind up in a home owner's own water supply.
25. *Pesticides.* The greatest danger of ingesting pesticides is with foods. Ensure that fresh foods are washed thoroughly, peeled, and outer leaves removed (as in lettuces and cabbages). Foods marked *certified organic* or grown by a home owner with organic compost are considered pretty much untainted.
26. *Right-to-know.* The term right-to-know refers to currently available information about a substancy product, of which can make the HES entrepreneur smart. Disclosure about product ingredients, hazardous chemicals used in manufacturing, emission statistics, etc., is mandatory and is reported to each state's Emergency Response Commission (ERC) and even local (county and municipality) Emergency Planning Committee (LEPC).

27. *Scouring powder.* Borax, table salt, and baking soda make effective scouring powders, or any commercial type that contains no coloring agents, chlorine, or detergents.
28. *Soaps.* White, unscented soaps are considered safest and probably the most economical, especially if allergies exist in the family.
29. *Solvents.* Chemical solvents, kerosene, and diesel fuel can be recycled by filtering them through a paper or cloth filter and reusing them. Gasoline is a no-no, however. It absorbs through the skin.
30. *Spray starch.* Some commercial ones can prove toxic. Homemade spray can be made by dissolving 2 tablespoons of plain cornstarch in a pint of cold water; shake well and store in a pump, spray bottle. For very delicate fabrics, use a package of unflavored gelatin.
31. *Ticks.* Similar to flea antidotes; see that listing. Check your doctor or medical clinic.
32. *Transmission fluid.* Don't fool with it. Deliver drained fluid, as is or mixed with waste motor oil, to a gas station that recycles the stuff properly.

Perhaps you will want to add a few more services of your own to this list and become a HES entrepreneur. Who knows what other services it can lead to in this service-oriented economy!

Home improvements: a billion-dollar opportunity

(See also Home Environment Specialist)

How to make millions of homes environment-friendly could be the most outstanding business opportunity of the remaining century. The housing market is subject to periodic ups and downs. Environmental problems and concerns, however, only go up. The development of an environmental consumer conscience is growing even faster. Ergo, opportunities exist for a multitude of trades and businesses to combine the needs of home owners with the demands for better living conditions. The equation adds up to a vast profit potential.

Home improvements often contain visible health benefits as well as demonstrable cash savings. The alert entrepreneur will be able to focus on these areas and show the prospective customer that the cost of home improvements can, in most cases, be amortized over a brief period—while the health and pleasure benefits are usually immediate. Let's roam around the typical house and point out areas of "home improvements" that will generate dual benefits:

Bedrooms

- Synthetic carpeting emits chemical vapors.

- Skylights are wonderful, where they are feasible, because they increase natural light (save electric lights) and, where they can be opened, allow fresh air to circulate.
- Improved insulation in walls and weather stripping around windows and doors can save over 30 percent of heat or air conditioning.
- No-iron sheets contain formaldehyde to which some people are dermatologically sensitive.
- A garbage recycling center should be part of any home. Find room for multiple bins for plastics, bottles, cans, and newspapers.

Laundry room

- If there is room, air dry laundry as much as feasible. It saves dryer energy and makes it smell better.
- At least 90 percent of hot water energy is wasted when laundering only on the "hot" cycle. Washing on cool or cold cycles saves electricity and dollars.
- Front-loading models use less hot water than top-loading washers.
- Clothes dryers can be made more economical if they have, or can be equipped with, a moisture-sensor—electric ignition—and a cool-down cycle.

Basement

- Where gray water tanks are permissible, these installations collect "gray water" from showers, baths, bathroom sinks, and laundry where they reuse it for nonpersonal uses, such as toilet flushing and landscape irrigation.
- Water pipes and the hot water tank should be wrapped in insulation for additional energy savings.
- Solar water heaters can save 75 percent of the energy used to heat bath and pool water.

From a pure profit perspective, some home improvements can return more dollar values than others. The three home improvements that return about 100 percent or more in home value appreciation are an interior facelift (average $5,000 and up), attic conversion ($10,000 and up), and basement conversion ($7,500 and up). An outside deck and inside fireplace are also high on the best-value-for-your-money improvements. Building in environmental concerns need not add to the cost, but only add additional advantages to the entrepreneur as well as the home owner.

Bathrooms

- Low-flush toilet bowls use one gallon of water or less, versus five or more gallons for ordinary toilets.
- Fluorescent fixtures and even supersaver incandescent bulbs use less electricity. One or two large lamps use less energy than a row of many smaller lamps.
- Aerators installed in faucets reduce water usage by at least 50 percent.
- Low-flow shower heads also reduce water consumption by at least 50 percent. In a family of four taking showers daily, this can save as much as 18,000 gallons of water annually plus the electricity or gas it takes to heat it.

Living room

- When repainting, use latex paints; oil-based paints contain pollution-causing toxins.
- Use windows and window glass that keeps the heat in during the winters and the heat out during the summers.
- Heavily lined drapes, insulating blinds, and shutters also help keep the home more comfortable and in intemperate weather, outside where it belongs.
- Use caution with wood stoves and open fireplaces that can cause substantial carbon dioxide air pollution. Investigate "safe" ones and replace old, dangerous ones.

Kitchen

- Don't install kitchen cabinets with materials that contain toxic formaldehyde. Specify solid wood or metal.
- While the trend is toward side-by-side refrigerators and automatic defrost models, both are far more energy-intensive than top freezer models (35 percent more efficient) or automatic defrost types.
- Make sure all hot water pipes are well insulated to deliver optimum energy saving.
- The best dishwashers have a short-cycle selector, an air-dry selector, and a built-in hot water heater.
- Compact fluorescent lights are energy savers, lasting 8 to 15 times as long as incandescent ones, although they initially cost more.
- If glue containing formaldehyde is used to lay kitchen flooring, home owners allergic to it could get long-term, low-level illnesses. Tile or concrete floors are best.

Remodeling for a kinder environment

The U.S. Department of Commerce estimated that remodeling projects cost nearly $80 billion in 1990. This indicates an increase of more than 6 percent a year for the past five years. It is a very big market with lots of opportunities—especially for honest purveyors of products and advice. Re-using or adapting existing materials saves considerable energy and timber and is a beneficial as well as a cost-effective recycling method.

In Massachusetts, a former architect became a consultant, not a contractor, who advised home owners and commercial customers on how to get the best value for their investment. He screened bids, arranged to buy supplies at discounts, and oversaw the project to a satisfactory completion. His charge was $65 and hour, and in 1990, he took in about $80,000—his third year in business. Ensuring that a remodeling project is done in an environmentally friendly manner is certainly a good position to be in.

Five years ago in New Orleans, an entrepreneur decided to become a "spaceman"—redesigning closets and building storage space for customers who needed extra room but did not want to sell their homes or offices. Redoing what already existed saved a great deal of energy and brought in revenue in excess of $100,000 last year.

In Atlanta, Georgia, the Electrolux Company formed the Home Services Alliance, then sold it in 1990 to a former construction supervisor. For a fee of $48 a year, the service acts as the superintendent on any home remodeling job. It checks out contractors, gets performance bonds, makes sure of prices and materials, and checks insurance and permits.

The Alliance is a worry-free service that has attracted 150,000-plus customers during the past few years. Over 170 contractors have also registered with the service, which doles out contracts to those best qualified. In turn, the contractor pays a graduated fee of 5 to 15 percent of the cost of the job to the service. During the past year, the Home Service Alliance awarded $2.5 million in contracts at an average gross income of $250,000. Such a service is also a great way of controlling the environmental quality of remodeling jobs. Like convenience foods, convenience remodeling is a business whose time has come.

Hot water heaters: getting into hot water for profit

Every day, 12,000 water heaters break. Replacing them is a $250 to $400 job. Equally bad, these heaters cannot be fixed economically nor recycled. They do not contain enough steel to make it worthwhile. Still worse, these bulky appliances wind up in some landfill where they take up a lot of increasingly scarce space.

There is now a preventative maintenance procedure developed by aerospace engineer Darrell Lemon of D.L. Enterprises, Longmont, Colorado. The company makes and sells a $40 replacement part called Flexi-Node. Sold through a network of dealers, the gadget can be installed simply with a set of instructions according to the developer. His dealers might even make an inspection—a great way to get one's foot in the door. With more than 90 million hot water heaters installed in businesses and homes, and heaters breaking down within 10 years, the market is tremendous.

Becoming a hot water heater inspector of any home or business that has such a heater 5 to 10 years old, could become a profitable business. With a little practice, the $40 Flexi-Node can be installed in 20 minutes. The landfill space thus saved—not counting the possible damage from flooding if a water heater breaks—is worthy of an environmental medal.

Hot water heaters powered by the sun

After 10 years in various areas of the solar heating field, Al C. Rich developed and patented the *Solar Skylight*. He claims that this innovation is a more attractive, more efficient, unbreakable solar water heater and backs his claim with a lifetime guarantee to the initial owner.

While the Reagan administration's solar heating subsidies were dismantled, it is likely that the Bush administration and Mideastern uncertainties will once again give solar heating a boost. Rich's American Solar Network of Herndon, Virginia has come on the market at the right time.

Domestic hot water heaters use more electricity than any other home appliance. The average annual electricity cost for a hot water heater servicing a family of four is close to $450. A solar heater can save as much as 85 percent of the electricity. From an environmental standpoint, it is informative to know that such a heater, supplying about 80 gallons of hot water a day, requires the energy equivalent of 11.4 barrels of oil per year—about the same as a midsized car driven 12,000 to 13,000 miles during the same period. In the case of oil or gas-fired hot water heaters, the hydrocarbon pollution is created at their domestic location, while electric heater pollution is created at the power generating plant.

Solar heating, as an important alternate energy source, clearly has a sunny future in America, from both the entrepreneurial as well as the environmental perspective.

Janitorial services: cleaning up

There are 35,000 janitorial services in the United States and 7,000 janitorial supply houses. This gigantic market for honest "environmental cleaning" companies or manufacturers of environmentally safe products has hardly been scratched. When you drive down any city street at night and look up, you will see thousands of windows lit—and tens of thousands of night workers cleaning up the mess left by day workers. That is your market, not counting the proliferation of maid services, internal maintenance crews in stores and factories, and households that would prefer environmentally benign products that pose no danger to children, pets, or sensitive hands.

Most offices have night crews that come in and empty waste baskets, run the vacuum over the carpeting, and perfunctorily run dust rags over open furniture areas. In some open floor areas, a Dust-Down type of product is used or hard floors are mopped.

These traditional cleaning methods do not tackle some of the more subtle, environmentally destructive accumulations of dirt, such as dust and grease that cling to ceilings and ceiling tile cracks and vents or the "electronic" dust in computer rooms, that is a by-product of heavy computer use.

Today, there are franchise operators that offer new cleanup techniques and devices that tackle this environmentally destructive dirt. One such company, Consol Carpet Cleaning, Stoughton, Massachusetts, franchises a central vacuum cleaning system that not only chases unusual dirt and dust accumulations, but does it without noise pollution through an out-of-office motor. Their concentration is on keeping computer rooms clean of damaging dust. Using hand-operated vacuum cleaners with micron filters, they are able to eliminate virtually 100 percent of dust and dirt. Periodic, antistatic treatments adds to the effectiveness—and the operator's profit.

Another new cleaning system concentrates on ceilings. It was developed by Ceiling Doctor International of Toronto, Canada in 1983 and expanded to the United States a few years later. Today, CDI has about 20 franchises in this country and has developed an annual volume of $5 million. So far, major customers who have had their ceilings cleaned professionally and their employees safeguarded from harmful aerial infection caused by dirt-harboring ceilings, include AT&T, McDonald units, doctors' offices, Allstate Insurance, and Merrill Lynch stock brokerages. The average cost of cleaning of ceilings runs 25¢ a square foot.

Landscaping: xeriscaping in the green

If xeriscaping is all Greek to you, be advised that *Xeros* is a word from the Greek meaning *dry*. Xeriscaping, therefore, is a way of landscaping in dry areas of the United States or when water shortages are endemic. It is a specialty that offers environmental enhancement and entrepreneurial challenges.

The first step in becoming a xeriscopic landscaper, therefore, is to analyze what plants grow well in a particular area and the soil you are analyzing. Using only plants that are native to the area can save as much as 50 percent in irrigation water. Applying xeriscaping principles to landscaping is aesthetically pleasing and ecologically sound—both emotionally and financially. The following are some of the considerations that need to be weighed in xeriscaping:

- Use drought-resistant plants like cacti, succulents, vines, ground cover, jasmine, bougainvillea, wisteria, daffodils, sweet alyssum, and other low-water use flowers and flowering shrubs.

- Specify grass species that are low-maintenance, such as Texas Bluegrass, which is watered about every four days, or Buffalograss, which needs water only every other week.
- Look into drip irrigation, heavier mulching of planting beds, and organic soil enhancement to allow for better water absorption and moisture retention.
- Check with the local agricultural extension agent or horticultural society. Send for a free *Xeriscape* brochure from the Texas Water Development Board, P.O. Box 13231, Capitol Station, Austin, TX 78711.

Landscaping the native way

The idea of landscaping with native plants is almost obvious: use locally available plants for residential and commercial landscaping and thus help restore and preserve the local ecology as well. Generally, this is what Ecohorizons Inc. of Goulds, Florida does.

Ecohorizons, a small community south of Miami on the east side of the Everglades has plants in profusion year-round. The company consults with home owners, commercial developers, and government agencies on the preservation and management of saltwater and freshwater wetlands, coastal areas, dunes, hammock (stands of old tropical trees and growth) in designing cost-efficient ecological landscapes—all utilizing primarily locally grown and available plants.

To augment Ecohorizon's work, the company affiliated itself with a local nursery that has been established for 15 years and has a good reputation in the environmental field. Educating established, as well as potential, customers in the value of restoration and protection of the environment is a large portion of the firm's marketing effort.

Plant doctor: clearing the air

A little environmental thinking can add value and profit to an existing plant business—or get a new plant business started with greater effectiveness. The secret is knowing what plants can do as air purifiers.

Forests absorb tons of harmful aerial pollutants to man and other living creatures. In the home or office, however, additional pollution makes life even more precarious.

There are ways of combatting "housatosis" of course. Open doors and windows when weather permits, use exhaust fans vented to the outside, especially in such areas as the bathroom, kitchen, and laundry. As part of your air conditioning installation, a heat recovery ventilator in the windows will also be helpful. Use of electronic air filters gets rid of

Antipollution Plants

Symptom	*Plant*
Formaldehyde pollution	philodendrons spider plants golden pothos snake plants
Benzene pollution	gerbera daisies chrysanthemums most other flowering plants
General air purification	reed palm English ivy peace lily mother-in-law's tongue Chinese evergreen dracaena aloe banana trees

Is there a doctor for the house? You bet!

microscopic particles in the air. However, whole-house units costs around $1,000, and room units run about $250 to $600. But then there are plants.

House plants can absorb many home pollutants. In fact, one single 10-to-12-inch potted plant will absorb aerial pollutants in a 100-square-foot area. If a 1,200-square-foot apartment, for instance, contains a dozen plants, the air will be considerably purer.

Light bulbs: how not to be ultra-violated

You wouldn't think that incandescent and fluorescent light bulbs could add to the environmental problems of our society. Billions of them are in use in every home, office, and workplace in the nation. Yet, many have built-in dangers from radioactivity, even though this danger is measured in nano or billionth of units.

How do you know which bulb gives off even minute amounts of radioactive rays? Each package says so; it is mandated by law. There is only one problem: the legend that tells consumers how many nanos of radioactivity the bulb emits is usually printed on the bottom of the package in such tiny print that it would be difficult to read it much less interpret it. What does "Contains 15 nano Ci Kr-85" mean? See what I mean?

Looking only at energy-savings, the new types of fluorescent bulbs, more compact than the old ones (and considerably more expensive upfront), are worth considering. They can fit into regular sockets and

> You can buy nuclear-free, ecologically safe, energy-saving bulbs from Ecoworks, an agency of Nuclear Free America. These bulbs are usually sold through environmentally minded mail order houses and cost about $1.50 each for the triple-life type that is rated to last 2,500 to 3,000 hours. Of course, wholesalers and mail order houses buy them in 30, twin-pack cartons considerably cheaper. The address is 2326 Pickwick Rd., Baltimore, MD 21207. And 10 percent of each order is donated to peace and environmental causes.

produce a warm, soft glow that is equal to the light from incandescent bulbs. The principal difference is that they last far longer, save 75 percent of electric energy and, by projection, eliminate the need for more power plants and the resultant pollution.

Let's take an example. The new compact fluorescent bulb can cost $15 (though no doubt the price will go down with increased production). In our house, we have 18 incandescent bulbs inside. That means a $270 investment, instead of a supply of regular bulbs costing perhaps $1 each. However, the new CFBs last 10 times as long; that equals to $180 in ordinary bulb purchases (18 bulbs @ $1 × 10 = $180). The new CFBs use only 1/4th of the energy. Normally, 18 incandescent bulbs use up about $80 worth of electricity in 9,000 hours; CFBs, during the same period, would cost $20 to burn. Consequently, after about 15,000 hours of cumulative burning, you are even. After that, the CFBs start paying *you* in energy savings.

Energy experts have figured out that if all Americans were to switch to CFBs, 40 large power plants could be retired and the scourge of acid rains and the greenhouse effect could be abated by a considerable margin.

Mr. Bulb is the name of a business started in Worcester, Massachusetts by entrepreneur-electrical parts distributor Alan Freidman. He operates a business that services small- to medium-sized commercial retailers via a brightly painted van run by a lighting specialist. The latter calls on 20 to 30 accounts daily, servicing a total of 800 customers. Replacing energy-saving fixtures and bulbs has become a sufficiently lucrative enough business that this entrepreneur is planning to franchise his business at the beginning of 1991.

Mail order: starting small, growing tall

Of the more than 10,000 listed mail order firms in the United States, a small but important segment specializes in a narrow or a very broad range of environmentally friendly merchandise. They range from a $75,000-a-year out-of-the-garage type of operation to one doing several million dollars annually with several very slick, full-color catalogs. It's

Environmentally Attuned Mail Order Companies

The Cotton Place, P.O. Box 59721D, Dallas, TX 75229

Cloth, fabrics, yarns, sheets, garments of natural cotton. Catalog available.

Ecco Bella, 6 Provost Sq., Suite 602, Caldwell, NJ 07006

Gift baskets, gourmet foods, recycled greeting cards, fragrances, Tom's of Maine cosmetics, natural pet products. All-natural botanical ingredients. 30-pg. color catalog.

Eco Design Co., 1365 Rufina Circle, Santa Fe, NM 87501

Livos paints and finishes, Alte Schule recycled and textured paper products, cedar colored pencils, natural art materials, polishes, pet products, body care products. Mostly from Germany. Catalog.

EcoPack Industries, 7859 S. 180 St., Kent, WA 98032

Extraordinary, springy packing material from unbleached paper. Literature available.

Earth Care Paper Inc., P.O. Box 7070, Madison, WI 53707

Greeting cards and paper goods, handsomely executed and all on recycled stock. Catalog includes informative "Environmental Agenda" columns. 40-page catalog.

From the Rain Forest, 8 E. 12 St. (#5), New York, NY 10003

Cashews and Brazil nuts from Brazil and dried fruits from Southeast Asia. No pesticides or preservatives used.

Globus Mercatus, Inc., 12 Central Ave., Cranford, NJ 07016

Flavo-It Natural Fruit Concentrates; appliances for making natural ice cream, yogurt, creams, and sodas. Did $200,000 in initial year in business (1989).

Healthy Kleaner, P.O. Box 4656, Boulder, CO 80306

Plant oil and grain cleaners to remove almost anything from skin and other surfaces. No animal testing; comes in recyclable plastic bottles.

N.E.E.D.S., 602 Nottingham Rd., Syracuse, NY 13224

National Ecological and Environmental Delivery System is a national shopping service for the ecologically sensitive person. Uses an 800 number to sell cosmetics, purifiers, and supplements. Catalog.

The Paragon Gifts, 89 Tom Harvey Rd., Westerly, RI 02891

A general mail order catalog house that is personally promoting environmental causes with recyclable paper and packaging. Features the Recycling Center ($45) and the Can Crusher ($22), Save the Earth books and net shopping bags.

Seventh Generation, Colchester, VT 05446

Jeffrey Hollender's and Alan Newman's well-established mail order company that sells from a colorful 32-page catalog (printed on recycled paper, of course), featuring more than 150 environmentally-benign products. Numerous ecological information panels (FYI) and pertinent promotions make this company a national standout.

Continued.

Upstate Eco-Logic, 29 Drake Rd., Newfield, NY 14867

Started selling "ecologically sound" products by mail in 1990 and ran up a $75,000 gross the first year. It can be done "when you do your homework, know your business, and know your customer," say the Elmira-area entrepreneurs.

Wireless, 274 Fillmore Ave. East, St. Paul, MN 55107

A slick, general mail order company that spotlights numerous unusual and popular environmental items in its colorful 48-page catalog. Planetary sweatshirts, earth sound recordings, and other "sound" offerings.

the one business an entrepreneur can start with very little money, limited space, and hardly more promotion than moderate classified ads in publications that cater to planetary causes.

Most mail order companies start with one or several related products, often the kind that they manufacture themselves. As these companies grow, they produce mailers or even catalogs of varying degrees of sophistication—expanding the list or quantity as business and fiscal liquidity warrant. When expanding, a mail order company will look for additional products, and even publish its need, in their promotional literature. Currently, there is still little ecologically friendly products, especially affordable ones that make them attractive buys. European marketers have found that customers are not willing to pay premium prices much in excess of "ordinary" merchandise—despite their avowed affinity for environmental causes.

Going into the mail order business seems easy. Certainly, it offers a great deal more flexibility to the beginning entrepreneur than a retail or manufacturing business does. It is a very tricky or deceptively easy business, however. For one, the percentage of returns from advertising and promotion is likely to be far lower than many entrepreneurs might have hoped. The medium, the size and type of list, the wording of the ad, the positioning with other ads, the competition, the popularity of the item advertised, and even the season or state of the economy are all factors that can affect results—and throw projections all out of kilter.

The cost of ordering, storing, and shipping must be added to the invariably high cost of promotion. Whether to buy and warehouse the merchandise or wait until orders come in is as important a gamble as proper pricing, competition, and exclusivity. The mail order business is a classic catch-22 business, and it requires expertise and luck.

Verena and Greg Sava of Birch River, West Virginia, started the Briar Run Farm where they keep, graze, and milk about 40 goats and make America's best chevre, fresh or aged. Chevre, as any Francophile knows, is goat cheese. And the latter, besides having a delicately pungent taste that is ambrosia to cheese lovers, contains half the fat of cream cheese, nearly half the calories and cholesterol, but nearly double the protein. It comes in a couple of dozen varieties and costs just a little more than ordinary stuff.

Briar Run Farm ships all over the United States to organic cheese lovers and gourmands who appreciate this blue-ribbon dairy specialty. Last year, the company shipped more than $60,000 worth of cheese—including a steady supply to the famed Greenbrier Resort Hotel nearby.

Marketing: quality and cause equal success

When it comes to unusual ice cream, superior ice cream, ballyhood ice cream, Ben & Jerry's Homemade Inc. are the winners, spoons down.

Ben & Jerry's has been churning out delicious frozen confections, as well as support for causes from family farms to rain forest preservation. The latest gambit is ice cream with jungle-grown nuts whose sale by the Brazilians is to be earmarked for the preservation of the Amazon rain forest and its dwindling aborigines.

Making Rain Forest Crunch will not spell instant commercial success for Ben & Jerry, but it points to the possibilities that alert entrepreneurs can find in environmental causes.

Some possible cynicism aside, such trendy marketing does aid the environment, as well as its entrepreneurial sponsor. But on the other hand . . .

Alcoholic beverages' salute to the environment

In addition to the Miller Brewing Company, which is using its beer brewing residue to enrich agricultural soil, the Japanese Suntory Brewery has become environmental. The $5 billion conglomerate has introduced a beer called *The Earth*, and if this is not enough, its slogan is "Suntory is thinking about the Earth," in Japanese and English. Suntory's market research showed that the ecology was becoming a hot issue in the Land of the Rising Sun. The company evidently is thinking that, with enough of the new beer sales, Japan could become the land of the rising Suntory.

Actually, thus far, the only environmentally friendly step has been Suntory's use of a stay-on closure instead of the usual metal flip-off, throwaway top. At least this innovation on Suntory bottles, and hopefully many other brands, will keep millions of metal flip-tops out of the landfills.

The Japan Union for Nature Conservations is politely disdainful, although the organization's spokesman admitted that the Suntory campaign won't do any harm and might even heighten environmental awareness among consumers.

New-trition: the way to health-y profits

Nutrition has gone through a revolution during the past few years. The seemingly sudden discovery of cholesterol has given a push to the trend of salads, fish, vegetable oils, and yogurt. The famed Gallup Poll put an amen on the trend by announcing that their national survey shows: 77 percent of men believe that the foods they eat directly affects their health; women, even more convincingly, said food affects their health by a resounding 88 percent.

Despite these figures, not all new-trition restaurants and groceries are successful. Too many variables can sink even the best concept—variables such as a bad location, inept management, inadequate financing, sanitary problems that cause municipal censure and, of course, just plain lousy food.

Given positive input by all the aforegoing operational facets, healthy food businesses have done very well and will continue to make money in an environmentally and nutritionally friendly context.

A&W Restaurants and Morrison's Cafeterias, just to mention two mass-feeding businesses, have added grilled chicken breasts to their menus, now one of the most popular items despite a premium price. Soup and salad restaurants are proliferating everywhere. Fast food purveyors feature the fact that they use only vegetable oils in deep-frying. Bresler's growth jumped 50 percent when they changed their name to include ice cream as well as yogurt.

Foods of the Future

Not only are dishes available in virtually every recipe that contains either low-fat ingredients or saturated fat substitutes, but food laboratories are working on the molecular structure of sugars and fats. The latest developments, available by the end of this decade, will be sugars and fats that taste just as good as the sweet and greasy stuff we enjoy today but that pass through the body undigested—and thus are benign to our vascular system.

Even fruits are being developed that are genetically resistant to insect pests, thus making heavy pesticide use unnecessary. Plastic film food wraps are being tested that can be safely eaten along with the foods they protect, keeping nonbiodegradable films out of our landfills. Cattle and other animals, as well as fish, will be raised to thrive on less body fat—and thus won't pass their fat on to us predatory Homo sapiens.

Baby food develops grown-up profits

In 1986, an angry mother who couldn't buy organic food for her baby decided to go into business. It happened to be the year Beechnut was indicted for fraud in selling sugar water as apple juice for babies and the year the Environmental Protection Agency banned the Alar preservative.

The start was as rough as any small underfinanced, inexperienced company's is. Sarah Redfield started Simply Pure Baby Foods in her Bangor, Maine home. She bought small, wide jars that were easy to get into and labels printed in nontoxic ink. Then she refinanced her home and bought two pieces of filling equipment; read up on the law (she was a graduate lawyer); got some help from friends in the Agriculture Department and the local university; and produced 2,000 cases of four types of baby foods. A Boston supermarket agreed to test market the baby foods, which sold well despite their higher cost. Evidently, mothers agreed that pesticide-free baby food was worth the difference.

The detailed and professional business plan she prepared did not cut too much ice at the bank, but served her well as a guide to developing her business. Said Redfield, "Your business plan doesn't have to be elaborate or done by a professional. As long as you have all the essential cost projections and marketing potential accurately stated, you should be okay."

In year three, she needed to expand and looked for more money—this time from "socially conscious venture capitalists." The results were the same: more frustration and no money. So she contacted all the friends she knew, told her story to them, and in a very short time, had the expansion funds available.

Simply Pure Foods Inc. now has more than 30 people on the payroll and is expanding beyond the borders of Maine and Massachusetts. The many organic farmers from whom she buys produce consider the company an excellent secondary market.

The baby food business is no kid stuff, as Sarah Redfield discovered, but faith in her idea, hard work, persistence, and a real consumer need for her product, spelled success, with a capital $.

Vitamins and health products: popping up prosperous

No matter how controversial our huge intake of vitamin products is with some medical circles, the public will respond to any siren call that promises better health and longer life. Taking vitamins is such a national obsession that thousands of shops have sprung up that specialize in vitamins, health potions and lotions, and organic food products.

General Nutrition Corporation (GNC) of Pittsburgh is one of the largest health-products companies with about 1,000 stores in most major shopping malls. While currently only about 100 stores are franchised, a total of 750 franchised stores are on the drawing board within the next four or five years.

It is a bit circuitous to say that such "health products" stores are environmentally friendly, but these establishments do purvey products designed to maintain and improve the health of their customers. They feature "natural" or organically grown products, and they offer entrepreneurs opportunities to make a living in an area of expected environment-friendly activity.

Studying human nutrition and purveying such knowledge along with the sale of vitamins, minerals, and other supplements and organically produced products, can make an entrepreneur in this field a more valuable citizen. Medical doctors emphasize sound nutrition only too rarely. This offers opportunities for the knowledgeable health products purveyor for as long as our current civilization exists.

Yogurt: from Russia with Glasnoszt

After Mike Smolyansky was in Chicago for a while, he thought fondly of a delicious and healthful yogurtlike drink that he and millions of other fellow Russians used to enjoy—kefir. If it was good for 260 million Russians, he thought, so it ought to be good for Amerikanskis.

In 1985, Smolyansky joined an American, George Allen, and formed Lifeway Foods Inc. They were able to scrape together $200,000 and opened a dairy plant in Skokie, Illinois to make the liquid milk culture similar to yogurt. After a few taste tests in local supermarkets, they found that sales in those stores typically rose about 75 percent, then leveled off to somewhat above prior sales. Still, these tests proved that Lifeway's kefir was a winner.

Their modest success did not go unnoticed by national yogurt producer Dannon, which came out with a liquid yogurt of their own. Still, Lifeway had a headstart and is now in a couple of dozen states. The heavy influx of Russian immigrants and other Europeans can only help Lifeway.

Investors must also think that kefir has potential in this country. In April 1988, Lifeway raised $600,000 in an initial public stock offering. When the stock was originally issued, it sold for $1 a share. The financial pages today indicate that Lifeway has indeed prospered. Nasdrovya!

Noise pollution: shhhhutting out cacophony

As our cities become larger and our society noisier, noise pollution enters as an environmental affront. Many concerned citizens are grumbling about trucks rumbling through main streets and motorcycles revving up engines as they speed by, or youngsters in open cars with rock music blaring to beat the band.

In the night clubs and on stages, rock bands have been vying not only for customers, but who can make the most noise. Heavy Metal aficionados claim that this is the only way to appreciate this pounding form of music. Hearing aid experts are rubbing their hands—when these musicians age, they will be customers for hearing aids. The audiences? They might not be far behind.

In Florida and some other areas where jet skis have made their appearance, local folks—including environmentalists, residents, resort owners, and Sunday boaters—have risen in protests to curtail the use of these noisy water toys.

Somewhere in all this noise pollution is a multiple opportunity to invent quieter engines, rechargeable electric ones that just hum instead of sputter, baffles that quiet down motorcycle and truck engines, and rock concerts that can only be heard through earphones.

Of course, this excludes ambulance, police and train sirens, and horns—and the person who insists on talking during a symphony concert.

Packaging: pollution solutions

In September 1990, Lever Brothers Co. opened a packaging laboratory in Owings Mills, Maryland, staffed by 40 engineers and scientists and headed by a vice president of packaging. This monumental event is perhaps the birth of Big Business's recognition that packaging can pollute the environment. Here, at this facility, research has begun for solutions to the solid waste problem by designing more environmentally benign packaging. The problems to which this Lever Brothers facility will try to find answers include using recycled plastics in laundry products bottles, reducing the amount of plastics that goes into containers (and thus into landfills), and using pigments and inks that do not contain heavy metal.

What some firms are doing

The ultimate in recyclable containers has been developed by a Buffalo, New York firm called the Kantain'R. It is a box that is designed especially for returnable aluminum cans, and holds 18, 36, or 48 cans. Since 44 percent of aluminum cans are still not being recycled, the market is immense.

Jim Walsh, the inventor and promoter, admits that the invention of the environmentally helpful container was relatively easy, but that marketing it will be difficult. Still, according to the *Beverage Industry Magazine*, 84 billion soda and beer cans made of aluminum were recycled in 1989, an increase of 11.5 percent over 1986. Walsh says that this adds up to $900 million in aluminum can recycling and is positive proof that his innovation, a 100 percent recyclable plastic (HDPE) container, will be a huge success. Innovative KantainR Inc. is at 136 Halwill Drive, Buffalo, NY 14226.

Much has been written about the pollution caused by nonbiodegradable grocery bags. At the Circle Supermarket in Carbondale, Colorado, Gary Miller, a 55-year-old checker, has asked the question, "paper or plastic" thousands of times. Reading about the nonbiodegradability of plastic and the fact that paper bags cost the lives of many trees, and remembering that most European shoppers use their own string bags, Miller created samples of the *Eco Sac*. This is a reinforced cotton canvas bag with handles, not unlike a carry-all bag marketed by the big L.L. Bean company.

The first 100 *Eco Sacs* Miller made sold for $13 at his supermarket. Utilizing production economies, he was able to sell following batches for under $10. His company then placed an order for 100 pieces for each of the chain's 37 stores. The "inventor" next hopes to convince his employer's parent company, the giant 1,200-store Kroger Co., to carry his *Eco Sacs* nationally.

Another company, JonEric Taylor and Leslie Koeper's Tread Lightly Traders, 2028 Murdstone Road, Pittsburgh, PA 15241, produces a 100 percent untreated cotton canvas grocery bag in three sizes and sells a set of four for $25. They also market them to wholesalers and fund-raising organizations, as well as a lot of sound environmental advice and information.

Among the several large supermarket chains that promote environmental consumer education, Giant Foods of Washington-Baltimore does a fine job of explaining to their customers the pros and cons of paper bags vs. plastic bags, biodegradable, photodegradable, recycling, and more. They also aid in the collection of used plastic bags and newspapers for recycling. Their booklet, "Paper or Plastic? There are no easy answers," is available from the Consumer Affairs Department, #597, Bldg. 1, P.O. Box 1804, Washington, DC 20013.

Opportunity

Packaging in foam cups is a no-no, according to environmental watchdogs. Throwaway foam cups have occasionally been touted as having "environmental advantages"—some are made without the use of the atmospheric pollutant chlorofluorocarbon. However, according to environmental experts, those that are made with the substitute, pentane, still create a low-level smog, or hydrochlorofluorocarbons, which does only slightly less damage to the ozone layer.

Other foam packaging is also a no-no. While Burger King started to eliminate environmentally harmful packaging in 1955. McDonald's, the world's largest fast-food purveyor, didn't jump on the ecology bandwagon until October 1990 when they announced that they would begin phasing out the familiar plastic foam boxes in its more than 8,000 restaurants in favor of more biodegradable paperboard containers. As the *New York Times* stated it, "The plastic McDonald's hamburger box may be on its way out, a packaging dinosaur that could not survive in a less wasteful age."

The entrepreneur who can invent or market a container that is environmentally safe and still has the properties of styrofoam—lightweight, insulating, inexpensive, and nonpolluting—will be lucrative indeed.

Packaging for a safer environment

For smaller entrepreneurs who are in the packaging field or are planning to enter it, a number of guidelines are available that offer great profit and public relations opportunities. These opportunities increase with the number of units sold, of course, such as in the food market. Here are just a few ideas:

- Provide a recycling machine that is actually a reverse vending machine. It accepts and processes glass containers, cans, and plastic bottles, and reduces them to compact residue. It can be programmed to return small amounts of cash to depositors. Source: Environmental Products Corporation, 11240 Waples Mill Road, Fairfax, VA 22030-6032.
- Set up separate bins outside of your establishment that will accept glass, aluminum cans, plastics, folded cardboard containers, and newspapers.
- Offer ideas on how containers can be used again. Run a contest to see what other uses your customers can come up with.
- Buy products that come in returnable containers. The extra effort involved in handling such containers means your customers must come again, hopefully to purchase other products.
- Promote products that are biodegradable, such as wax paper, glass casseroles, and cloth products and explain their advantages to your customers in advertising and displays.
- Feature products and produce that customers can wrap in paper or design wraps of degradable papers rather than nonbiodegradable plastics.
- Test packaging for weight-holding ability and produce it as light as feasible. Use recycled products if they are cost comparable.
- Package and promote larger quantity units that cut down the amount of packaging required and promote their use along environmental lines.
- Increase display space and bulky packaging departments and provide biodegradable containers in small as well as large sizes. Many bulk products departments only offer large bags, which might be wasteful.
- In designing packaging, make sure the product can either be recycled or biodegraded, that the ink used on packaging is nontoxic, and the paper stock itself is not bleached with poisonous dioxin (a highly carcinogenic chemical).

Paints and coatings for the environmental age

The new Clean Air Act, once in full force, will affect hundreds of everyday products, especially in the paint and coatings industry.

Paint

Virtually all house paints contain various amounts of environmentally damaging hydrocarbons. Paint companies have gradually improved their lines with the emphasis on water-based paint, usually in response to tightening federal and state regulations. Water-based paints generally have five to ten times less hydrocarbon content than oil-based paints and give off fewer fumes. Pittsburgh Paint's PPG Industries is one major supplier that has been testing low-hydrocarbon paints for several years. Other technicians are working on ways to filter air in spray-painting rooms in auto factories and repainting shops.

Spray painting

It is perhaps ironic that Union Carbide Corp., working with Johns Hopkins University chemical engineers, has developed a painting process that could potentially reduce paint spray pollution by 30 to 70 percent.

Spray painting normally contributes to aerial pollution problems because of the various solvents used to dilute the paint so that it can be sprayed. The volatile organic compounds that make up the solvents evaporate during the painting process, react with nitrogen in the atmosphere, and contribute to smog.

The new spray paint technique replaces about 65 percent of the solvent with nonpolluting, recycled carbon dioxide. Moreover, this new technique is inexpensive, nontoxic, and nonpolluting. The Environmental Protection Agency is now researching the method and, should it be approved as expected, the safer manufacturing spray paint could be incorporated into the new Clean Air Act. If that happens, many licensing opportunities for smaller paint manufacturers should open up—as well as a substantial increase in the value of Union Carbide stock.

Curing the varnish process

Varnish is most often used in wood finishes. It seals wood and gives it a luxuriant shine, making it last longer and of course, sell better. Unfortunately, varnish emits harmful pollutants into the air.

Like paint, however, the varnishing process has likewise undergone some serious environmental research. Battelle Memorial Institute of Columbus, Ohio has developed a process in which varnish is cured in a matter of seconds rather than over a matter of hours in open air or fume-generating ovens.

This new Battelle development allows varnish to be cured by intense ultraviolet light in a matter of seconds. With this method, most of the pollutants get locked into the finish instead of escaping into the atmosphere.

Should this quick-drying, nonpolluting drying method be incorporated into the Clean Air Act, it will create a boom for the chemical finishing industry, the furniture business, and the environment.

Metal coating

More eclectic opportunities exist in the airplane and metal coating field. Working with ultraviolet light, new coatings have been developed for metal surface finishing that prevent corrosion by adhering tightly to the surfaces to which they are applied. These new coatings do not shrink as they dry, preventing them from cracking as old coatings often did. Moreover, they have no solvent emissions—the very problem that have brought the old procedures under scrutiny of health organizations and bureaus. One resource for the new shrinkless coatings is Epolin, Inc., 358 Adams Street, Newark, New Jersey 07105.

The paint industry, as in many others, opportunities exist in new, environmentally friendly innovations. Here, as elsewhere, necessity became the proverbial mother of invention.

Pizza purveyors preserve power

The distance between a pizza baker and energy conservation seems to be an astronomical one. In Hollywood, California, however, a company named El Centro Foods has worked out a plan to save one half of the considerable gas that its 62-unit chain of pizza parlors uses.

El Centro's chain, Pizza Man He Delivers, uses natural gas ovens to bake the tens of thousands of pizza pies. Bills from the Southern California Gas Company started mounting along with inflation and the price of oil. The company decided to do something about it, both to conserve energy and, in their own self-interest, expenses.

With the utility company's help, the pizza chain began to install more cost-effective ovens that had a lower BTU rating. Even though the cost was almost $10,000 per installed new unit, utility bills were cut in half immediately.

The transformation to greater energy efficiency gave El Centro the impetus to start franchising its chain—all with the new equipment. "We felt a responsibility to make sure we were respectful of our natural resources," said vice president James Dalton.

And what's good for pizza bakers is probably good for any other high-energy user.

Politics: if you're seeing red, switch to green

In "pre-Anschluss," "green" politicians have already been elected by their active, vociferous and politically astute environmental constituents. In the United States, we have not yet reached this level of political environmental clout but it appears to be coming.

In Nebraska, a man named Hugh Kaufman is assistant to the director of the Hazardous Site Control Division at the EPA. He has been the environmental activists' inside man in the EPA bureaucracy and blew the whistle on his own agency when he felt the EPA was not doing enough to control and eliminate toxic waste sites.

In 1990, he ran for the office of representative of the village board of Nora, Nebraska. This crossroad on the map has only 21 registered voters, but Kaufman was elected and is now perhaps the only elected official of the "Green" party—though he ran as a write-in candidate, which is allowed under the Federal Hatch Act.

One reason that Nora, Nebraska is important is that a proposed nuclear waste dump could be located nearby. The citizens of Nora and their elected official, Hugh Kaufman, are against it.

It is only a tiny opening in the traditional bipartisan political picture of the United States, but it is worthwhile keeping an eye on it. The next environmental entrepreneur might make a run for a bigger political plum . . .

Pollution fighters: entrepreneurial opportunities

One futurist who tried to describe what life in the next millennium might be like, opined that the American government will have to start dismantling the very big cities, spread them around the less developed countryside, limit cities to a maximum of 300,000 population, and have city planners start from scratch, developing environmentally planned model cities. Meanwhile, what can the small to moderate developer, planner, and landscaper do to help the environment recover—while helping himself? The following are six areas that the Natural Resources Defense Council has pointed out as needing attention:

1. *Open area.* Build or develop undeveloped areas in housing and commercial developments. Such open spaces will absorb rainwater and recharge the groundwater supply. Vegetation keeps the water from washing salt and pollutants into streams.
2. *Pavers.* Be conscious of the problems that vast cement or macadamed areas create. Such sections cease to absorb rainwater and help to erode banks and flood rivers, adding silt to water that kills aquatic grasses, fish, and shellfish. Perforated brick paving in vital areas can help to prevent these problems and add profits to the installers.
3. *Organic waste pollution.* The biggest culprit of farm and suburban residents, includes garbage, animal droppings, leaves, and grass clippings. Proper disposals by waste management organizations and educating householders in proper disposal methods are two businesses that can benefit in the process.
4. *Fertilizer control.* Lawn and garden fertilizers and organic wastes produce nitrogen and phosphorous as they decompose. These chemicals cause algae to grow in profusion, robbing water

of its oxygen and harming aquatic life. Producing better fertilizers, educating householders, and handling decomposition matter in an environment-friendly manner are triple opportunities for enterprising entrepreneurs.

5. *Zinc pollution.* Zinc pollution can usually be traced to aging pipes and gutters, as well as runoff, weathering, and abrasion of galvanized iron and steel roofs and sidings. Mixing such runoffs with groundwater is dangerous, and besides, it doesn't taste good. Repainting old iron and steel with safer coatings and replacing aging gutters and pipes are just two profitable areas for alert contractors.
6. *Copper pollution.* Most copper pollution is caused by worn automobile brakes than by worn pipes and fittings, although both contribute to water purification problems. Being alert to these problems can increase business for plumbers and brake repair shops.

Printing for the environment

Ken Abraham wanted to run a printing business with a difference. He uses only recycled papers and recycled plastics. His ace salesperson is an ugly little green monster whom he called RecycleSaurous.

Abraham's company, Creative Printing and Publishing, 712 N. Highway 17-92, Longwood, Florida 32750, is the place RecycleSaurous and his pals live. The child-oriented trademark appears on two sizes of comic and activity books, bookmarks, buttons, bumper stickers, games and toys, T-shirts and caps, stationery, and other specialties. An 800 number (800-780-4447) connects this mid-Florida firm with the world.

While Creative sells the environment with educational plugs in their catalog, they also do a good job in selling themselves to Americans all over the country who appreciate Creative's customized merchandising and environmentally friendly efforts.

Proving that small journals can also be a big help in promoting the environment while also profiting from the ecological movement, Patricia Poore created *Garbage, The Practical Journal for the Environment* in 1989. This no-nonsense, slick periodical was preceded by *In Business*, an idea-full journal out of Emmaus, Pennsylvania, published by Jerome Goldstein for the past four years, principally for a loyal cadre of entrepreneurs.

A new start-up journal called *Design Spirit* was launched by Suzanne Koblentz-Goodman, an architect, and her husband Jordan Goodman, a reporter at *Money* magazine, in December 1989. Its purpose is also linked to environmental trends by showing architects, builders, and sculptors how to use "compassion, wisdom and transcendence . . . spirit-uplifting and ecologically-sound designs.

This printer's award-winning program offers inspiration to the 50,000 printers—10,000 of whom are small, instant-print shops—who might be looking for an angle that will make them stand out from the crowd. Volunteering for the environment is a plus that is possibly good for the soul. In Abraham's case, it is also good for his business.

Recycling

The recycling industry offers enormous potential for the green entrepreneur, especially as Americans become increasingly aware of reducing their own household wastes and corporate consciousness becomes more marked. Even in those recycling industries such as paper, aluminum, and newspaper that have been around for some time, the full market potential has yet to be realized.

Aluminum recycling

The members of the Lions Club in Bethany and Rehoboth, Delaware get up early every Saturday morning to make the rounds of assigned collection points. They pick up previously planted plastic containers filled with empty aluminum beer and soda cans. From each of the hundreds of such drops, the bags go to central collection containers where trucks pick them up and haul them to an aluminum recycling depot.

It is rare that you can find a stray aluminum can lying around. Everybody is suddenly so ecology-conscious that gas station and grocery operators, schools, and tavern keepers work hard to show their good citizenship. Best yet, the empty cans bring 35 cents a pound.

Aluminum recycling goes on all over the country—sponsored primarily by Anheuser-Busch breweries. In 1989, 475 million pounds of aluminum cans were collected and resold to aluminum fabricators for a total return of $166 million.

Because making new aluminum uses seven kilowatt hours in electricity, recycling old cans is far cheaper. The nationwide effort is so important to the environment and makes so much economic sense, that Anheuser-Busch has acquired its own aluminum recycling plant, Container Recovery Corporation.

Cartridge recycling

For the past several years, laser printers in thousands of offices are using cartridges. These cartridges contain a drum fueled by chemical toners that are used in the printing process. The average laser printer owner uses about 10 to 12 cartridges each year. The empty ones, like so many other peripheral office equipment, becomes garbage and winds up in an already-crowded landfill. Until about five years ago, that is.

A few entrepreneurs back around 1985 saw an opportunity to recharge those $120 replacement cartridges. It did not take too much ingenuity to do the job—just some pliers, a large container, a vacuum, and a bulk supply of toner.

The process of recycling laser printing cartridges, however, is cumbersome and inefficient. So another entrepreneur, Vern Williamson, formed Controlled Environments, Inc. in Las Vegas and developed a more efficient manufacturing device called "Cartridge Master" that sells for $2,495. It can refill and recycle up to 75 cartridges a day and eliminates the pollution that accompanies the old hand operation.

Recharged cartridges can sell for $49 to $59 each. More than 5,000 small companies already are in the business, but the way laser printers are coming onto the market, lots more opportunities for entrepreneurs are opening up. A trade group has also been formed to guide the new industry and assure quality control—the International Computer Products Recycling Association.

Paper recycling

The Undaunted Recycler is a typical example of a small recycler, blending environmental enthusiasm with entrepreneurial aspirations. Dan Hayden started this business in 1988, picking up "presorted, high-grade paper," as well as offering file removal and shredding of confidential papers. As a sideline, it seems, he offers copying services and unbleached, 100 percent recycled paper for sale.

While this entrepreneur offers free collections of computer printouts, white and colored as well as NCR papers, he is also educating his customers and potential resources with a steady stream of environmental data.

Publicity, free and generous, has helped the Undaunted Recycler get his business off the ground with very little promotion expense. Recy-

In the paper-selling business, either as producer, wholesaler, or retailer, you won't want to forget this easy-sell pitch:

- It takes 500,000 trees, a whole forest, to produce enough newsprint to publish the nation's newspapers on a single Sunday.
- Today, only about 30 percent of the 80+ million tons of paper come from recycled stock; the rest comes from virgin wood pulp.
- Each one of us in the United States "consumes" 580 pounds of paper annually—a total of 50 million tons—in noncommercial use. That adds up to 850 million trees.
- Those of us who work in offices are responsible for an average of 180 pounds of high-grade recyclable paper thrown away each year.
- Each ton of paper that is recycled saves 380 gallons of oil needed to produce virgin paper.
- Considering the growing scarcity of landfill land, it is good to remember that each ton of recycled paper saves three cubic yards of landfill space. And since 50 million tons of waste paper are accumulated by consumers alone, a national recycling program could save 150 million cubic yards of precious landfill space.

cling is a hot topic and newspapers and local publications seem to like to write about it. "Dan the Recycling Man," "Business Paper Recycling Efforts Gain Acceptance," "Undaunted Recycler Fills Niche," and "Wastebuster" are the headlines that have appeared in both local and national publications.

Hayden started the business with about $2,000 for a used truck, some fiberboard receptacles, and a hand truck. He has 125 clients and collects up to 10 tons of paper each month. He estimates that about 80 percent of what he now collects for recycling used to wind up in local landfills. Both customers and the local county commissioners are happy about Hayden's "creative approach."

Printing, shredding, and supplies

Three related enterprises are prime sources for recycling of good papers: (1) 10,000 instant-print shops and as many as 40,000 other printers; (2) thousands of paper shredders, including those working for banks, brokerages, and government agencies; and (3) 15,000 office supply firms. Entrepreneurs in these fields can profitably set up paper recycling procedures. Most of all, they can use and promote recycled paper stock by educating their employees and customers. In most cases, the switch to sorting trash paper and using recycled paper means developing an environmental consciousness—breaking wasteful habits.

Three fast-printing franchise companies are making productive efforts to switch their thousands of franchisees and hundreds of thousands of customers to the recycling mode. These are Sir Speedy of Laguna Hills, California; Alphagraphics of Tucson, Arizona; and PIP Printing of Agoura Hills, California.

Data Destruction Services developed a mobile paper shredding service that has been recycling processed papers for more than 2,000 clients for nearly a decade.

One large office supply firm in Cleveland, ProForma, is working at educating customers in the switch to more recycled paper products.

Newspaper recycling

It seems like an unlikely way of grossing in excess of half a million dollars a year, but that's what John Lohr's American Cellulose Co. did in 1989. The plant, located in Minonk, Illinois, a crossroads of less than 3,000 people, a little north of Bloomington. Here's how he did it:

In 1976, Lohr went to work for American Cellulose as a salesman. The company was environmentally conscious, and this appealed to him. In true Horatio Alger fashion, he worked hard and saw an opportunity to buy the business in 1987 for $150,000.

The company has collected old newspapers from municipalities, nonprofit groups, and churches, and recycled them into light, fluffy insulating material that can be sprayed between walls. The business is acutely affected by the ups and downs of energy prices, so Lohr diversified.

The sterile, recycled newspapers were reduced to a mulch, bagged in 25-pound sacks, and sold as a less expensive alternative to wood chips. From bagged chips it was a small step to manufacturing those popular compressed logs that burn in fireplaces for about three hours. The newspaper mulch is combined with decontaminated, recycled motor oil—thus further eliminating a waste product and putting it to new purpose. These rolls are sold under the name of *Oregon Fire Logs*.

In March 1989, a new product was added: pulverized newsprint was mixed with water and grass seeds and packed into 33-pound bales. They are sold to pavers and municipalities who want to encourage ground vegetation in barren areas, either to hold the earth down or as roadbeds.

American Cellulose, by saving thousands of tons of old newspapers from cluttering up crowded landfills, thus became one of Illinois' prime environmentalists. In fact, his county gave Lohr the 1989 Business of the Year Award. State Governor Jim Thompson added to the company's kudos by awarding it the 1989 Illinois Governor's Recycling Award.

Environmental concerns continue, however. The next product American Cellulose will introduce is Enviro-Dry, a product that will also use recycled newspapers to sop up spilled oil in automotive and machine shops.

The problem with old newspapers is complicated, To re-use newsprint, it must first be bleached of its printed surfaces. There are not enough de-inking mills around, and new methods must be innovated to make this process practical. Lt. Governor Mark Singel of Pennsylvania commented that this need will create added opportunities and foresees the newspaper de-inking business as a very lucrative one.

Plastic recycling

One of the problems in recycling waste plastics has always been the recovery cost and finding practical uses for the resultant end product. PolySource Mid-Atlantic, 7120 Golden Ring Road, Baltimore, MD 21221, seems to have closed the loop.

PolySource is one of the new post-consumer plastics recycling companies. Their resources are local communities, package manufacturers, and industrial and environmental groups who help bring recovered waste plastics back into the packaging stream.

The result of the company's recycling effort is an extruded pellet or flake that can be re-used as a superior packing material, or as raw material for remanufactured plastic products. An example is many of the new liquid soap and cleanser bottles used by Procter & Gamble in such products as Spic 'n Span, Tide, Cheer, and Downy.

Plastic recycling is still a business of the future. Much of its success depends on consumer education. Jean Statler of the Washington-based Council for Solid Waste Solutions, said there is a market for recycled plastics because they are generally cheaper than virgin or new plastics.

"Plastics recycling is going to be a big business," stated Statler. "It's one of the most recyclable materials. It's not a question of demand for recycled plastics, it's a question of supply . . . Consumers aren't separating plastics from their other garbage and the municipalities aren't collecting it."

Future success—and this is where *green* entrepreneurs' ingenuity comes in—will hinge on the cooperation of participating municipalities and their citizens. The latter must get used to separating used plastics from other wastes. By doing this, both taxpayers and municipalities will save a lot of money—primarily on vacating the need to find costly new landfills.

Entrepreneurs who are entering this new field of plastic waste recycling should be aware that an analysis of wastewater from a plastics remanufacturing process showed that pollutant levels were well within the Environmental Protection Agency's standards and local standards (in Baltimore)—another plus for plastics recycling. No air is emitted from the recycling process, because the only heat involved softens the shredded plastic so that it can be extruded and sliced into pellets or flakes. The recycled plastic now being used for nonpersonal bottles by Procter & Gamble can also be used to manufacture trash bags, pipes, rain gutters, and downspouts.

In Maryland, a state tax-exempt revenue bond of $9,970 million was floated to construct a plastics processing plant in Sparrows Point. It will be the first plant of its kind in the United States and is based on a working prototype plant in Coburg, Germany.

Recycling in the park

A company as big as Dow Chemical Co. or Huntsman Chemical Corp. can recycle on a big scale. These companies made arrangements with the U.S. Department of the Interior, which runs the nation's public parks, to set up recycling bins that they will erect, collect, distribute and recycle. These companies will try to convince the millions of people who visit national parks that their paper, glass, and plastic wastes should go into properly marked bins.

There are thousands of smaller parks across the nation that could be "patrolled" for waste products, and local waste collectors and recyclers could literally "clean up." Where states and municipalities fall short—either through lack of funding, imagination, or both—private entrepreneurs can take up the cause for both the environment and their own profit. As Dow's president Frank Popoff put it, "Demonstrating the importance and feasibility of recycling is the best way we know to encourage a recycling ethic."

Recycling on the beach

Like parks, beaches are also prone to suffer from people pollution. The bane of almost every beach resort is the amount of trash that accumulates on beaches and boardwalks, despite warning signs, fines, and waste receptacles. This is not merely a malaise of U.S. coastal beaches, but also of the supposedly idyllic beaches of the Caribbean islands.

Ocean City, Maryland is one resort that has launched a coordinated clean up and recycling operation. During the beginning of summer 1990, Ocean City officials had placed 150 extra barrels along the ocean front for recycling aluminum cans and glass. Other containers were set up to receive plastic products for recycling.

Maryland's state conservation director was drafted to work with local resort personnel. Getting his cooperation was not difficult because Maryland Governor William Donald Schaefer was the first one to come up with the recycling idea in April 1990 during the dedication of a Baltimore recycling plant. Ocean City resort officials have been scurrying ever since to implement the popular governor's plan. They even had the Governor to come to Ocean City at the beginning of the season to kick off the campaign.

To make beach visitors aware of Ocean City's recycling efforts, a series of billboards were strategically placed along the approach highway, Route 50. The town's public works crews pick up the empty aluminum cans and turned them over to a regional recycling center for reprocessing.

Perhaps inspired by Ocean City's example, the Lions Club in Bethany Beach, Delaware, a few miles up the road, organized an aluminum can collection and recycling plan through local media and schools.

The utilization of municipal and civic club facilities can provide entrepreneurial opportunities for wide-awake recyclers.

Slot machines arm for recycling

A Swiss inventor came up with a machine that is fed not by quarters but by aluminum soda and beer cans. While it crushes the cans for eventual recycling, it spits out one of five possible prizes as a reward—coupons for sodas, or discounts on grocery products or on other products distributed by one of the five sponsors.

In the Spring of 1990, the first several *LuckyCan* crushing machines were installed in a number of Florida supermarkets for on-site testing. *LuckyCan* operates like a slot machine, however. Not every can depositor wins. Those who don't still get a card that has imprinted on it an environmental slogan about recycling.

The Swiss-made, Las Vegas-inspired contraption is distributed in the United States by The Howard Marlboro Group, the sales promotion division of the international advertising agency Saatchi & Saatchi.

Everybody but everybody is getting on the environmental recycling bandwagon and joining the Green Revolution.

Site surveys with an environmental twist

Today, locating the right site for a factory, warehouse, or other commercial structure, or even buying an existing one, is fraught with environmental implications that hitherto were never considered. Real estate professionals and others planning to enter the real estate business, will have to be acquainted with environmental impact studies (EIS).

The Environmental Protection Agency, various state licensing bureaus, and even banks who invest in properties and could be held liable for environmental pollution within or under structures mortgaged by them, are all concerned with EIS.

An environmental impact study analyzes the impact of a proposed plant or building on the quality of life in an area. Agencies typically require the study to cover such topics as impact on existing transportation facilities, energy requirements, water and sewage treatment needs, effects on natural plant life and wildlife, and water, air, and noise pollution.

Individual state requirements sometimes are even stricter than federal ones. In New England, for example, the construction of an oil refinery was aborted by Maine and New Hampshire. Delaware has a Delaware Coastal Zone Act that prevents heavy industry from locating within two miles of its 115-mile coastline. A new jetport that was planned to be located on the edge of the Everglades west of Miami was blocked by pressure from environmental groups fearful of further endangering the delicate ecological balance of the 5,000-square-mile Everglades area. There are many other considerations that site finders or realtors will need to weigh. Among these, are:

- Proximity to markets
- Proximity to raw materials
- Availability of skilled personnel
- Availability of labor
- Access roads
- Railway or water transportation
- Public transportation
- Tax structure
- Land costs
- Construction costs
- Competition
- Community living conditions
- Local zoning restrictions

With all these "normal" considerations, the environmental ones take on equal or even greater importance: supplies of water, availability of energy at acceptable costs, and lack of hazardous conditions that might be present, like ghosts in a haunted house, way down in the earth.

Other factors can also affect whether a piece of property is a good buy—not all of them chemical or included in an EIS. Many areas of the country have endangered trees and other plant life, which could prevent you from building on a site. Land located in a flood plain, greenbelt, swamp, or wetland could have serious development problems. Finally, habitats of endangered species cannot be developed.

Consequently, a cottage industry of topographers, horticulturists, and other plant and land experts specializing in environmental concerns for commercial real estate developers is great, and it goes without saying that those real estate professionals who are knowledgeable about environmental ramifications will have the extra edge.

Travel: see the sights to save them

Ecotourism is a new concept suddenly discovered and being promoted by a number of travel agencies. Developing respect and concern for the environment is what ecotourism is all about. The best ecotours invite travelers to experience beautiful and unique natural habitats, emphasize care for the ecosystem, and even produce economic benefits that encourage conservation.

Among those companies and institutions that are promoting ecological tours are Virgin Airlines of Great Britain, which plants a tree for each passenger flying between Los Angeles and London; Overseas Adventure Travel of Cambridge, Massachusetts; International Expeditions of Birmingham, Alabama; Outdoor Adventure River Specialists, Angels Camp, California; Jackson Hole Land Trust, Wyoming; and the National Audoubon Society.

Other, nonprofit organizations active in the ecotourism business include Earth Watch, Cultural Survival, The earth Preservation Fund, and the Center for Responsible Tourism, headquartered at 2 Kensington Road, San Anselmo, California 94960.

Some of the tour operators donate portions of their gross income to maintenance of nature trails, preservation of endangered species, and contributions to environmental organizations. Peter A. Berle, president of the 560,000-member Audoubon Society, confirmed that "tourism and the environment can peacefully co-exist. We know from long experience with travel programs . . . that travel can indeed be a strong economic and educational force for environmental protection."

One travel agency, Travel Associates, Riverdale, Illinois, books only "eco-tourism" travel to all seven continents. They state that "eco-tourism will be the fastest growing field of travel in the next 10 years."

Water marketing: money, money everywhere

In various parts of the country, water shortages make life for city dwellers precarious. Life-as-usual often has to be curtailed because sources of water supplies that most of us take for granted—primarily lakes—are either drying up or are being polluted beyond usability.

In the summer of 1990, Lake Mono, in the Sierra Nevadas of California, was drying up. Not only was the entire ecosystem endangered, but far-away Los Angeles, whose millions of inhabitants depended on the waters of Lake Mono for about 17 percent of their water supply, was told to start conserving water.

This near-tragedy (or deprivation, at least) did not go unnoticed. Entrepreneurs jumped into the breach where federal and state officials were waffling in red tape, indecision, and internecine competition, and violà—the water marketer was born.

Investigation into the water shortage revealed this to be a two-sided problem. One was the rampant growth of the cities and the flagrant abuse of water use. The other was the high use of fresh water by the state's farmers for irrigation during frequent dry spells. In fact, an estimated 85 percent of the waters went to irrigate huge farms, using existing inefficient methods of irrigation.

Paid out of a combination of municipal and corporate funds the water marketers went to work. They visited farmers in high water use areas to persuade them to use less wasteful methods—such as piped in water and sprinkler systems, similar to those used in the Middle East's arid areas like Israel. In extreme cases, deals were worked out to temporarily let acreage remain unused and, therefore, unwatered. This nonuse compensation was paid for out of water revenue funds.

Everywhere the entrepreneur has an opportunity. Utilizing existing water supplies more effectively is certainly one of the major priorities.

Car washes: getting more mileage from water

The average car wash uses about 10 gallons of water per vehicle. So-called "gypsy" car washes that operate from portable tents, garages, or driveways use as much as 50 gallons per job.

The way professional car washes conserve water—and that is the majority of them, and in South Florida, about 90 percent of those checked—is with a recycling mechanism. Runoff water is collected in one tank, processed through a filter system, and pumped into another storage tank. The reclaimed water is used over again as the next vehicle goes over the trestle.

Specialty Systems, manufacturers of the installation in Minneapolis, say that the same water can be used several times if the sediment is allowed to settle and the runoff filtered. "Car wash water does not have to be beautiful spring water. All it has to be is wet," stated the company's spokesman. These self-serve systems cost about $20,000, though some car wash systems, especially the deluxe "brushless" ones, can go up to $40,000 per installation. The latter can use, on average, 50 gallons of recycled water and 15 gallons of fresh water for rinsing.

Some car washes have promoted themselves as "water-saving" but their performance does not always keep up with their promises. Any entrepreneur going into this business should make sure that the manufacturer's claims are indeed provable by his water bills because there are many variables involved.

Toilets: too much water down the drain

Between four and five gallons of water is used each time you depress the handle on an ordinary toilet bowl. Water conservationists have recommended putting bricks into the tank or anchoring filled, plastic water bottles in the tank. Both techniques displace regular water volume and thus, the toilet flushes a gallon or so less than normally. To launch a Home Efficiency Consultant enterprise, these little tricks are good to know. However, in Creswell, Oregon is the Con-Tech Industries firm that makes *Future Flush*. This patented device is for people who forget to conserve water, and that is most of us. It features a double-handled toilet flusher, allowing for shorter flushes for paper and liquid wastes; another handle for flushing through more solid wastes. Because some states already mandate water-conserving products, manufacturers of toilet equipment will want to investigate this twin-handled water-miser.

If you are remodeling or building a home, consider an ultra low-flush toilet. It is pressurized and empties the tank with about one gallon of water or even with compressed air. This ULF keeps water trapped at the maximum pressure in the line until it is released and squirts into the bowl. Local plumbing supply stores should be familiar with it.

Imagine if the average large city were to convert its toilet tanks to water-saving ones—if only by installing a weighted plastic bottle that displaces one gallon of water. Since each toilet is flushed about eight times a day, eight gallons of pure water would be saved daily. In a year, that adds up to nearly 3,000 gallons of water. If a city like Kansas City, Missouri, with a population of nearly half a million people and about 160,000 housing units and probably at least 200,000 toilets, were to pass an ordinance mandating one-gallon water displacement bottles in the city's toilet tanks, 600 million gallons of water could be conserved.

What is the cost of this avalanche of water in one single year? How much would 200,000 one-gallon bottles equipped with tight tops and rustproof handles that hook over toilet tank edges cost? These WDCs (water displacement containers) could be filled with recycled water, too, to enhance their environmentally friendly character even further. Distribution could be handled through the local water company's meter readers, or through drug, food, and hardware stores. What a boon for drought-prone areas! What a boon for a local entrepreneur!

Water: pure, potable, and profitable

Residential water purification is approaching the $2 billion mark—and that is a sizable market. This includes $139.95 attachments, $10 five-gallon bottles, fancy-label bottled waters, and $29.95 cartridge filter buckets.

The environmental advantage of this burgeoning industry might be in doubt, but it is certainly an opportunity for entrepreneurs. If the claims of the sellers are even close to reality, then imbibing filtered waters is assuredly a health measure. If nothing else, these waters taste better than many of the doctored municipal waters out of the tap.

Microreactors to the Rescue

It sounds like something out of a science fiction movie, but it might soon become reality. Polymer molecules can be dissolved in water, absorb solar energy, break down even the most serious pollutants, and transform them into harmless chemicals.

These water-soluble synthetic polymers are called photoenzymes. They are being tested at a University of Toronto laboratory by chemistry professor Dr. James E. Guillet. Results of the tests thus far make these photoenzymes effective for cleaning water contaminated with PCBs, polynuclear aromatics—which are carcinogens—and other harmful pollutants.

As in any relatively new and fast-growing business, some charlatans see a health trend as an opportunity for a passel of fast bucks and wild claims. Some Texas purveyors of bottled waters were in that league and had their corporate knuckles rapped by the government.

It can be envisioned that selling purified waters is not only selling better taste and health benefits, but reducing local water needs, and as such, this industry is an environmentally friendly one. And besides, projections over the next five years are that the industry's sales will more than double, to a hefty $3.8 billion. It is indeed a good time for informed entrepreneurs to get into the water . . .

8

Managing by environment: how your business can make a difference

MANAGING BY ENVIRONMENT ACTUALLY OFFERS SOME OF THIS DECADE'S most promising and satisfying business opportunities. It is a simple matter of adopting a mind-frame: we will do nothing that hurts the world we live in; if there are problems, then we must devise solutions; if we cannot devise solutions by ourselves, we will join others in effecting satisfactory solutions. But how do you organize your business to be environmentally friendly? There are three approaches you can take depending on the size of the business:

1. Large enterprises can change all or part of their operation to be environmentally compatible, partially because of corporate conscience and partly because it makes good marketing and public relations sense. Perhaps they can implement recycling plans or maybe develop new environmentally safe products.
2. Smaller companies can gradually change their merchandise or service procedure so that they are compatible with the environment, including possible changes in suppliers, retraining employees, and adjusting their pricing to accommodate the new policy.
3. An entrepreneur, convinced that doing business in concert with the environment is compatible with his philosophy and the reality of a

future world in which he and his children will have to live, sets up his new business entirely along environment-friendly lines.

Big business is making the change

There can be no doubt that many big businesses have redirected some of their operations to be more environmentally attuned. Everyday, new "converts" have begun to implement earth-saving philosophies simply because being good to the globe makes sound marketing sense. As more and more consumers come to recognize the value of environment-friendly marketing, the more public relations value will be mined by those companies that have made serious efforts to reverse their former laissez-faire attitudes.

A triad of recent converts includes the state of Texas, which has had the highest air pollution rate of any state in the Union. Because agriculture is a major part of the Texas gross product and a major source of its pollution, its emphasis on foods grown without chemical fertilizers and pesticides is certainly to be lauded. Jim Hightower, the Texas Commissioner of Agriculture, said that nonchemical crops are "growing dramatically. This is not a market fueled by a bunch of hippies studying their mantras. It is mainstream. This market is a big part of the future of American agriculture."

The gigantic 3M Corporation of Minneapolis instituted a "Pollution Prevention Pays" campaign that promotes in-house recycling of industrial products. 3M has publicly admitted that this campaign has saved the company millions of dollars.

No newcomer to the recycling scene, the American Telephone & Telegraph Co. has satisfactorily conducted an internal paper recycling program. In 1988 alone, the company announced a $365,000 profit on this phase of its huge operation and a total of more than $1 million since AT&T started the program.

Keeping the earth greener, it appears, keeps corporate treasuries greener.

Public skepticism: right or wrong?

Unfortunately, bigness appears to be incompatible with environmental conscience in the public's mind. Still, it must be said that large companies are trying to help clean up the ecological mess they have helped to create. Their very bigness and multitudes of departments and middle-managements often makes them stumble over their own feet, however.

England, which is a highly concentrated industrial nation but only 1/37th the size of the United States, can be dissected as a test. This is exactly what *The Economist*, a British weekly business publication, did for a number of weeks. The publication alerted its international readers to the fact that it is time to learn about environmental issues. *The Economist* reported on the sad state of our ecology and on what some large companies are already doing to reverse the disastrous trends. Finally, it also noted that some corporations ran into unexpected cynicism on the part of the public.

Quoting a report from Lloyds Bank, the periodical stated that the bank found that sponsoring environmental projects did not seem to raise its credibility with consumers. They reported that the public is skeptical of any large organization that suddenly jumps on the environmental bandwagon, particularly those that have little direct contact with the environment. *The Economist* correctly noted, as many U.S. companies have also discovered, that enterprises are fearful of environmentalists' backlash if they get their corporate empathies wrong—in which case the corporate sponsors, instead of earning "greenie points," earn more public scorn.

Environmental campaign basics

It is only natural that a number of environmental groups have noted naive or exaggerated claims by some manufacturers. In fact, six such groups have called for a boycott of suspect products. Furthermore, a number of states have passed laws to control frivolous claims as well as to encourage more research and use of degradable containers.

Large and small companies need to realize two phases in any environment campaign:

1. Concerns that need to be addressed must include the entire operation—employee attitudes, vehicle maintenance, plant operation, product content, packaging, recyclability, and public perception.
2. Have a contingency public relations plan handy in case anything goes wrong, even if it is beyond company control.

In this supersensitive area of environmentalism, it is all too easy to be caught with egg on one's corporate face. The best way to prevent such embarrassment is to keep management on every level aware. They must really understand what environmental concerns are all about, why we must preserve our environment, and why we must be scrupulously honest in our statements, advertising, and labeling.

Today, the public is more aware of environmental problems and issues than ever before. No longer can you fool some of the people some of the time. Even if you can get away with environmental abuses in the short-run, some alert competitor or bureaucrat will catch you in a lie or even an omission. Today, jumping on the "Managing by Environment" bandwagon is not enough. Entrepreneurs should consider how to position their product environmentally, how to turn old liabilities into assets, deal with regulators, create new products or services, and understand the psychology of packaging.

Positioning your product environmentally

To position your product environmentally, you must produce an environmentally sound product or change an existing one to conform with the best environment-friendly requirements—and then blow your horn for all it's worth. Sometimes, a government agency supports such efforts, such as in Canada, Germany, and Japan. This step is not always easy to attain and is subject to official scrutiny, testing, and certification. Instead, utilize your industry group or trade association to obtain such recognition. Welcome established environmental groups to assure themselves that your products are indeed all that you claim for them—and perhaps their "seal of approval" will follow. Like the well-known Good Housekeeping Seal of Approval, or the approval emblem for a medical or dental society, the environmental imprimatur is the OK seal of the nineties.

Sows ears into gold purses

This is an allegorical way of saying that companies can turn old liabilities into new assets. So many large and small companies have proven that, intelligently handled, environment-friendly changes can lead to the preservation and development of greater market shares and higher profits. The only enemy of old-line companies is their resistance to change. There are alternatives to tropical oils, chlorofluorocarbons, impermeable plastics, and heavy metals. Any environment-friendly change from these bugaboos of industrialization is cause for jubilation and good public relations. Even additional profits on occasion.

Living with the regulators

In our bureaucratic, litigious society, new regulations that hope to turn back the threats of engulfing pollution are coming out on an assembly line. Should an eager legislator or one of his minions discover an environmental concern, you can bet your last tax dollar that a new environmental regulation is not far behind. As I have previously pointed out, even competitors in the same business will not miss an opportunity to snitch on you. In such an atmosphere, the entrepreneur must not only weigh and abide by existing regulations, but anticipate them. Will there be a tax on carbon products, such as fossil fuels, that is intended to be earmarked for pollution control? Will coal-fired power plants be subject to a pollution tax? Will your new vehicles be gas guzzlers in order to get to the next red light a little faster? Expect to pay an extra tax for that privilege. Will a little cheaper construction on your next building or home cost more in the long run because of inefficient insulation, groundwater pollution, or toxic emission? Anticipate these environmental problems, both physical and legal. The public will pay just a little more if you know how to sell these advantages intelligently.

Creating new products and services

The new decade and the millennium beyond will require that business-as-usual be relegated to the dynasty of the dinosaurs. We have new problems in the world and new needs, and consequently, we need new solutions—even if they will only counterbalance the infractions of past generations. There are opportunities everywhere; one need only to look. The media, schools, trade and professional associations, cultural institutions, and our fellow-man are all brimming with ideas waiting to be implemented. Big companies call this R&D and invest a definite proportion of their profits in researching and developing advanced concepts and products. Small entrepreneurs can do no less in proportion to their abilities.

Understanding psychology and packaging

You need not be a Freud to reason and figure out what your customers or clients want. When you make changes or introduce a new concept or product, do it positively and with conviction and enthusiasm. When you approach a dog with soft words and confidence, he will soon stop growling (probably this applies to everyone except mail carriers!). When you sell an ounce of perfume that smells very much like a dozen others, it's the package that initially attracts the buyer. A beautiful table setting, an attractively arranged platter, and a pleasing aroma will make a meal seem to taste much better. Jimmie Carter, in a classic, oft-repeated tale, tried to sell home heating conservation to a spoiled American public by advocating wearing sweaters in a cooler house. Americans were unwilling to make these sacrifices. They might have cooperated if he had sold them the idea of getting an insulation survey and insulating the home better in order to save energy. As old Elmer Wheeler, the promotion guru, intoned back in the thirties, don't sell the steak, sell the sizzle.

Getting help

You are not alone if you plan to switch over to environment-friendly products or services or start new ones. You have plenty of company. In this book, you'll find more than two dozen organizations that can help (see Appendix C). Uncle Sam's Environmental Protection Agency has dozens of specialists offering advice and free information, which is covered in chapter 9. Also, chambers of commerce, local economic development agencies, trade, and professional societies all have experienced advisors. The Small Business Administration and the National Institute of Standards and Technology are two other federal entities that can help.

Small Business Administration

The U.S. Small Business Administration is also in the business of helping entrepreneurs preserve and improve the environment. The newly created

Pollution Control Loan Program provides long-term financing to help small businesses plan, design, and install pollution-controlled facilities. Such environmentally helpful facilities might include recycling, treatment, liquid, and solid wastes containment, as well as aerial pollution. For further information on the program, contact the SBA's Pollution Control Loan Program at (202) 205-6552.

National Institute of Standards and Technology

Entrepreneurs who have developed unusual devices or processes in the environmental field, devices or processes they feel might be patentable, can get a free boost from a government agency. For energy-related technology, contact the U.S. Department of Commerce's National Institute of Standards and Technology. This agency evaluates energy-related inventions and recommends those it considers practical to the Department of Energy. The DOE, in turn, can provide financial assistance. Grants for commercially useful inventions average $72,000.

Details about patent information can be obtained from a related agency called NTIS. The free guide's name is *National Technical Information Service's Products and Services 1990 Catalog*. The NTIS request number is (703)487-4650, while the NIST number is (301)975-5500. If you prefer to write NIST, they are in Building 411, Room A115, Gaithersburg, MD 20899.

No matter what your line of business, you are not alone. Management by Environment is a growing concern to everyone. There is no holding back the entrepreneur who knows how to express himself.

9

The Environmental Protection Agency: help or hindrance?

IN 1970, THE U.S. CONGRESS ESTABLISHED THE ENVIRONMENTAL PROTECTION Agency and budgeted it for more than $3.5 billion. In short order, the EPA acquired more than 9,000 employees, as well as an often unjust reputation of being antibusiness. It isn't. It is pro-environment—an entity shared by both entrepreneurs and non-entrepreneurs. As a matter of fact, the EPA has an ombudsman to act as a buffer between the bureaucracy and the businessperson. The current ombudsman, Karen V. Brown, can be reached at (703)557-2027 or 1938 or the toll-free EPA hotline: 800-368-5888. You can also write to 401 M Street, SW, Washington, DC 20460.

Services and programs

The EPA has literally hundreds of experts in various fields, both business and scientific. If you have any doubt about the environment-friendly nature of your business or need to know about the government regulations of any business, scientific, or academic activity, there is a specialist you can reach through the ombudsman's office or the general information number.

Experts are available in the following areas: drinking water, noise, radiation, R&D environmental engineering and technology, R&D processes and effects, solid wastes, toxic substances and pesticides, and water quality.

The INFOTERRA, a "worldwide information" program is for businesspeople who want environmental data base information on any government agency, private organization, university, individual, or any one of the 115 participating countries. Direct questions can be sent to the EPA's ombudsman's office.

Data on the toxicity of more than 56,000 chemicals and compounds is also stored at the EPA through a program called RTECS (Registry of Toxic Effects of Chemical Substances).

The EPA also maintains an environmental constituency specialist. This fancy title is the office of the official in charge of keeping track of all known conferences, meetings, workshops, seminars, and conventions that have any dealing with the government's actions regarding the environment.

On occasion, grants are given to private enterprises ranging from less than $5,000 to many millions. Grant areas cover air pollution controls, wastewater treatment facilities, construction management, solid waste management, hazardous substances remedial projects, marine bays and estuaries protection, pesticide control research and enforcement, water system safety, and pollutants research on both humans and animals.

The EPA study "Impacts of Environmental Regulations on Small Business" (EPA 230-09-88-039) details environmental data on sensitive industries, including:

- Electroplating
- Wood preserving
- Pesticide formulation and packaging
- Farm supply stores
- Interstate trucking
- Gasoline service stations
- Dry cleaning
- Photofinishing laboratories
- Water supply

Entrepreneurs who wish to pursue more information from the EPA, might want to send for their 12-page bulletin, *Information for Small Business.* It contains hundreds of pamphlets and publications, most of them available free from Karen V. Brown, U.S. Environmental Protection Agency, 401 M Street, SW, Room A-149 C, Washington, DC 20460. The above bulletin also includes a detailed order form. Finally, the EPA has several regional small business liaisons that can also help (see Fig. 9-1).

Name and Address	*Official-EPA Title*
Lester Sutton Small Business Liaison Region I, U.S. Environmental Protection Agency New England Regional Laboratory 60 Westview Avenue Lexington, MA 02173 (Phone: (617)860-4355)	Special Assist to Regional Administrator
Donna Giannotti Small Business Liaison Region II, U.S. Environmental Protection Agency 26 Federal Plaza, Room #505 New York, NY 10278 (Phone: (212)264-4711)	Environmental Engineer Air & Env. Applications Sect. Administration Branch Office of Policy & Management
Richard Kampf Small Business Liaison Region III, U.S. Environmental Protection Agency 841 Chestnut Street Philadelphia, PA 19107 (Phone: (215)597-9817)	State Liaison Officer Office of Congressional and Intergovernment
Annette Hill Small Business Liaison Region IV, U.S. Environmental Protection Agency 345 Courtland Street, N.E. Atlanta, GA 30365 (Phone: (404)347-7109)	Program Analyst Office of Policy Planning and Evaluation
Margaret McCue Small Business Liaison Region V, U.S. Environmental Protection Agency Attn: 5-PA-14 230 S. Dearborn Street Chicago, IL 60604 (Phone: (312)353-2072)	Acting Director of Public Affairs
Phillip A. Charles Small Business Liaison Region VI, U.S. Environmental Protection Agency 1445 Ross Avenue Dallas, TX 75202 (Phone: (214)655-2200)	Director of External Affair
Charles Hensley Small Business Liaison Region VII, U.S. Environmental Protection Agency 25 Funston Road Kansas City, KS 66115 (Phone: (913)236-3881)	Deputy Director Environmental Services Div.
Charles Stevens Small Business Liaison Region VIII, U.S. Environmental	Environmental Protection Specialist Office of External Affairs

Fig. 9-1. U.S. Environmental Protection Agency regional small business liasons.

Fig. 9-1. Continued.

Name and Address	*Official-EPA Title*
Protection Agency Attn: Mail Code OEA 999 18th Street, Suite 500 Denver, CO 80202-2405 (Phone: (303)294-1111)	
Marsha Harris Small Business Liaison Region IX, U.S. Environmental Protection Agency 215 Freemont Street San Francisco, CA 94105 (Phone: (415) 556-6429)	Staff Assistant to Assistant Regional Administrator
Mary Neilson Small Business Liaison Region X, U.S. Environmental Protection Agency 1200 Sixth Avenue Seattle, WA 98101 (Phone: (206) 442-4280)	Constituency Coordinator (Freedom of Information/ Education Specialist)

The EPA, if you should need them in your entrepreneurial efforts, can be a friend in need (which a tax-paid public servant should be) or a pain in the neck. It is up to John Doe to make sure he gets his dough's worth, counting the change with kid gloves rather than iron fists. And if all else fails, remember that one of the entrepreneur's best ombudsman is your local congressional representative.

Voices from the other side

In any country as vast as the United States and any bureaucracy as immense as the U.S. government, confusion and red tape are bound to make their appearance. This is more likely when one considers the many self-centered interests represented by 50 states, hundreds of legislators elected by parochial interests, and at least two political party views, and several dozen societies that each have their own agendas for the environment.

The nature of government in the nation's capital makes for a huge, costly, and often contradictory, cadre of lobbyists—"specialists"—who have "connections" and for sums varying from mere expenses to several thousand dollars a week, will help unravel the Potomac red tape. A system not very dissimilar exists within the federal government as well.

The Environmental Protection Agency, despite its goodwill, is not always able to act unilaterally. Each division of the government has its own priorities, which it wants to fulfill come hell or high water. Sometimes, the high water is mighty murky.

Take, for instance, the number of hazardous waste sites that have been identified by the EPA and that, under existing rules, must be cleaned up within a prescribed period. Out of 4,615 hazardous waste sites on the cleanup list, the U.S. government owns 338. For almost two years, the EPA has been waiting for rules, which must come from the Office of Budget and Management (OBM), under which it can proceed to get the job mandated by congress done. And for two years, the OMB has shuffled these rules from desk to desk. While the cleanup of the 338 hazardous government-owned sites is mighty costly, it will be a lot costlier to clean them up in the future.

The same problem is passed on to private businesses. They, too, must operate under "corrective action rules," which were enacted under the federal Resource Conservation and Recovery Act. The law specifies how the job must be done but the rules are to be issued by the OMB. Until those rules are issued, it would be a waste of time and money for any business to do more than a superficial job. Meanwhile, as of the summer of 1990, none of the 4,615 hazardous waste sites have been cleaned up.

The problem remains but the opportunities are still there—somewhere ahead on the other side of the political fence, wrapped in red tape. Eventually, it is felt, that that wall, like the Berlin Wall, must come tumbling down. It will be interesting to see whether this will happen during the tenure of the current "environmental president."

And where does the U.S. Congress stand, or sit, in this mess? While the congress passed the law and presumably allocated the funds for the cleanup, it might have been diverted by another equally big cleanup—the savings and loan bail out. If the congress missed an opportunity for realistic action, it was because the lawmakers naively passed the cleanup bill without deadlines. They trusted the EPA to do the right thing, but they did not figure on the OMB. And while the review of the many-sided problem is being discussed and investigated, nobody but nobody is talking.

In another case, the banning of the pesticide chemical *alar* produced a similar cornucopia of confusion. The Natural Resources Defense Council issued studies in the Spring of 1989 that claimed alar was indeed carcinogenic. Uniroyal Chemical, alar's manufacturer, it must be said, took the chemical off the market when the announcement was made. The company's action came more as the result of the public's outcry than federal action, however.

Michael McCloskey, chairman of the 550,000-member Sierra Club, reacted strongly to the alar danger and the EPA's inertia. "We put all our eggs in the basket of legislation," he stated, "and increasingly it looks like a lot of input and very little output." "We've passed a lot of good laws," he continued. "We've passed some of them twice. What's the sense in going back and passing them a third and fourth time? The agencies aren't going to enforce them anyway."

McCloskey pointed out that the EPA had been charged with developing hundreds of safe standards for pesticides, going back as far as 1972. "But so far," he said, "over 90 percent of those standards have yet to be developed."

What the Sierra Club and other environmental organizations are continuing to do, is not to rely totally on legislative action, but to use consumer muscle to force safer standards. "There's a danger of getting too many actions going at once and having too much competition for people's attention." McCloskey added, "Federal regulators can't enforce anti-pollution laws, because there has been an explosion of issues that have to be decided on a political basis . . . they cost powerful companies lots of money. Bureaucrats are not equipped to do that. So they surround themselves with a maze of regulations designed to insulate the bureaucrats from political pressure. It's not working."

In conclusion, the head of the Sierra Club told an environmental symposium of the American Association for the Advancement of Science, "The time has come for 'green consumerism'—to use the power of the public to tell companies that we expect them to be good stewards of the environment."

The companies, it appears, are responding—faster than the elected government.

EPA goes to Europe

Laying the groundwork for possible entrepreneurial participation in the cleaning up of the European environment, is the EPA-sponsored International Environment Committee (IEC). A committee composed of 31 academic and corporate representatives was formed to explore markets for environmental technology, financing solutions, technology transfer mechanisms, and an international environmental information network.

In Budapest, Hungary, an independent, nonprofit center is being established that addresses problems throughout East Europe. The EPA, U.S.A.I.D. section of the State Department and Western European governments and organizations are cooperating. A seminar on environmental education was organized in Prague, Czechoslovakia by U.S. representatives and local counterparts and students in September 1990. On September 25–28, 1990, the U.S. embassies in Paris, Stockholm, and Helsinki sponsored a series of environmental seminars that brought together European and American interests.

The IEC committee director is Jan McAlpine, U.S. EPA, Office of Cooperative Environmental Management, A-101F6, 401 M Street, SW, Washington, DC 20460, (202)382-2477. The EPA's detailed FACT SHEET on this project is shown in Fig. 9-2.

The Clean Air Act

Somebody, possibly in jest, has called the Clean Air Act the growth industry of the 1990s. The Act will pump as much as $100 million into the U.S. economy in various ways. Rather than regarding it as a government burden imposed upon the backs of the poor taxpayers, the Act will actually open up many avenues of business on all levels of every industry.

The Regional Environmental Center for Central and Eastern Europe

What is the center?

The Regional Environmental Center, scheduled to open September 1990, is an independent, nonprofit organization, established to address environmental challenges that are common to Central and Eastern Europe.

The Center was proposed in July 1989 by President Bush as part of a program of environmental initiatives to help the newly emerging democracies of Eastern and Central Europe. The United States Environmental Protection Agency has been working together with the Republic of Hungary as well as the other founders, the European Economic Community and Austria, to establish the Center. Governments and organizations from around the world are indicating strong support for the establishment and work of the Center.

Where will the center be located?

The Center will be located in Budapest, Hungary, and will serve the Central and Eastern European Region. The Center will be housed in a 200-year-old silk mill, located in Obuda. The address is: 1 Miklós teŕ, 1035 Budapest, Hungary.

What is the mission of the center?

The mission of the Regional Environmental Center is to help the citizens, environmental organizations, the private sector and government agencies of Central and Eastern Europe address the natural resource and environmental problems threatening prospects for ecologically sustainable development in the Region, thereby fostering improvement, prevention and protection activities.

How will the center be organized?

The Center will operate under the general policy direction of an international Board of Trustees. The Board will be comprised of the founders of the Center, environmental experts, government officials, business and industry leaders, representatives of foundations, academics, environmental and other non-governmental organizations.

The founding governments will select the Executive Director and the Program Manager to oversee the operation of the Center. The staff, selected by the Executive Director, will consist primarily of Central and Eastern Europeans. Some Westerners will be used in a transitional capacity to train Central and Eastern Europeans in environmental planning and management skills.

What is the role of the executive director?

The Executive Director will be responsible for the overall program, financial and operational management of the Center under the policy and budget direction of the Board of Trustees.

What is the role of the program manager?

The Program Manager will report to the Executive Director and is responsible for coordination and

Fig. 9-2. EPA fact sheet for The Regional Environmental Center for Central and Eastern Europe.

Fig. 9-2. Continued.

implementation of programs and activities approved by the Board of Trustees, and supervises the programming staff.

What is the role of the operations manager?

The Operations Manager reports to the Executive Director and is in charge of all of the administrative management activities of the Center, including financial and endowment management, staffing and personnel, contract supervision, logistics, facilities management, and supervision of the operations staff.

What services will be available?

The Center will perform four major functions: data collection and dissemination; education and outreach; clearinghouse; and institutional development. There is a preliminary agreement that the Center will focus its initial work on programs concerning three critical areas: 1) addressing the major impacts on health caused by environmental degradation, 2) encouraging less reliance on pollution control in favor of more cost effective pollution prevention, and 3) facilitating the adoption of an energy efficiency policy that advocates reducing energy use which will result in a decrease in such pollution.

EPA regulatory assistance for small business

EPA's Asbestos and small business ombudsman

The Asbestos and Small Business Ombudsman will make sure that your problems and suggestions are heard and you are fairly treated. The Ombudsman:

- gives you easier access to the Agency;
- helps you comply with EPA's regulations;
- investigates and resolves your problems and disputes with the Agency; and
- increases EPA's responsiveness and sensitivity to your needs.
- Provides, or gets, answers to technical questions.

The Ombudsman closely follows the development and status of EPA's policies and regulations affecting small businesses, solicits your ideas on how to improve them, and makes sure that your ideas are carefully considered.

You can save time and money (and much frustration) by calling the Ombudsman at 1-800-368-5888. In DC and Virginia: (703) 557-1938.

A member of the Ombudsman's staff will answer between 8:30 a.m. and 4:30 p.m. EDT. Message-recording devices for calls during non-business hours and overload periods are provided.

You may also write the Asbestos and Small Business Ombudsman at the U.S. Environmental Protection Agency, (A-149C), 401 M Street, S.W., Washington, DC 20460, or visit the Ombudsman when in the Nation's Capitol.

If you're a small businessman, the Environmental Protection Agency

Fig. 9-2. Continued.

knows you don't have time to be president, general manager, shipping agent, advertising executive, chief bookkeeper, plant mechanic, and truck repairman, while discovering, interpreting, and meeting environmental requirements.

Yet the law requires you to comply and assumes you know what to do. You want to do what's right, but you don't have a staff of engineers or other technicians, nor hundreds of hours to spend on air or water pollution or on waste problems.

What's "right" for the large companies of this world is often "not right"—and—shouldn't be right—for you. This pamphlet shows what we're doing to bridge this gap.

The Office of the Ombudsman keeps up to date on all environmental matters of concern to small business. To achieve this, a network of Ombudsman Liaisons is active in each of the components of EPA. Through this network, the entire array of diversified technical expertise within the Agency contributes to the Ombudsman's response to specific inquiries from small business. The Ombudsman maintains alertness to small business concerns through hotline inquiries and through close contact with small business "umbrella" organizations, industry-specific trade associations, and relevant governmental entities. Additionally, assistance is provided through development and distribution of publications and audiovisual materials, as well as participation in small business seminars, workshops, and conventions.

Asbestos technical and regulatory information

Information concerning asbestos abatement problems is available also through toll-free calls to (800) 368-5888, or in DC and VA (703) 557-1938.

The questions answered covers those about the Asbestos Emergency Response Act of (AHERA) of 1986, and other questions requiring technical response within EPA.

Pollution control financing

A compilation of information covering Assistance Programs for Pollution Control Financing at Federal, state, and local levels is available on request from the EPA Small Business Ombudsman. This information identifies and describes in detail possible sources of financing. One source is the Pollution Control Financing Guarantee program of the Small Business Administration (SBA), specially designed to help small business firms secure long-term credit for pollution control needs. For information on this program, call (202) 653-2548, or write to: Pollution Control Financing Staff, Small Business Administration, 1441 L Street, N.W., Suite 808, Washington, DC 20416.

Other available assistance

ISCA-toxic substances

To help businesses comply with the Toxic Substances Control Act (TSCA), a variety of services are provided, including:

- seminars, publications, audiovisual materials;

Fig. 9-2. Continued.

- advice on premanufacturer notifications; and
- regulatory aid through a small-chemical company liaison based in Washington, DC.

The TSCA Assistance Office telephone number is (202) 554-1404. You may also write to the TSCA Assistance Office, Office of Toxic Substances, U.S. Environmental Protection Agency (TS-799) 401 M Street, S.W., Washington, DC 20460.

RCRA and CERCLA (or Superfund)
Solid and hazardous waste

Toll-free telephone service is available for information on the Resource Conservation and Recovery Act (RCRA) and the Comprehensive Environmental Response, Compensation, and Liability Act (CERCLA) better known as Superfund. RCRA authorizes the Agency to regulate hazardous wastes. CERCLA, or Superfund, establishes an emergency response mechanism for hazardous substances that may threaten human health and the environment. For related assistance, call (800) 424-9346 (in Washington, DC: (202) 382-3000).

National pesticide telecommunication network

The National Pesticide Telecommunication Network Hotline provides medical personnel and others with help in coping with pesticide poisonings, and also provides pesticide products information. The toll-free number is (800) 858-7378; in Texas call (806) 743-3091.

Chemical emergency preparedness, emergency planning, and community right-to-know

This Hotline provides response to questions, literature, and other assistance concerning community preparedness for coping with chemical accidents, as well as information covering the Community Right-to-Know component and the toxic chemical inventory release reporting requirements of the Superfund Amendments and Reauthorization Act (SARA). The toll-free number is (800) 535-0202; in the District of Columbia and Alaska call (202) 479-2449. This Hotline is an information resource service, not an emergency response number.

National response center

The National Response Center for Oil and Hazardous Chemical Spills is operated by the United States Coast Guard. The center should be contacted for initiating emergency mobilization of resources for coping with spillage and accidental chemical releases. The toll-free number is (800) 424-8802; in the District of Columbia call (202) 426-2675.

Safe drinking water

The Hotline (800) 426-4791 in DC (202) 382-5533) provides information on the Safe Drinking Water Act (SDWA), policy, technical and regulatory questions.

EPA opportunities for small business

The Office of Small and Disadvantaged Business Utilization (OSDBU) makes sure that a fair

Fig. 9-2. Continued.

proportion of business opportunities resulting from EPA's direct procurement and grant activity go to small business, disadvantaged firms, women's business enterprises, and firms located in labor surplus areas.

For information concerning these opportunities, you may call (703) 557-7777, or write to OSDBU, U.S. Environmental Protection Agency, (A-149C), 401 M Street, S.W., Washington, DC 20460.

The Clean Air Act, like its sister law, the Clean Water Act, is no panacea that will magically eliminate environmental impurities and problems. At best, it is evolutionary, not revolutionary. It is flawed with all the loopholes of ingenious and devious man.

The Act's manager, the Environmental Protection Agency, is often burdened with its own inefficiencies and political hamstrings. The EPA's biggest problem is that the Act demands implementation by individual states. That alone dilutes the Act's value by nearly 50 percent. Each state has its own politics, leadership, special needs, funding capabilities, pressures, and aspirations. Somewhere along the line, some of the Act's provisions fall through the cracks of time and inertia.

The Clean Air Act was enacted in 1955. That revelation alone is a wonderment! Our legislators had the foresight more than 35 years ago to dictate to industry: clean up your act or we'll do it for you.

Of course, laws are only as effective as the people who vote and provide the funds. And so the Act floundered around through numerous amendments in 1963, 1965, 1967, 1970, 1977, and once more in 1990 when President Bush signed into law the most comprehensive of all Clean Air Acts on November 15.

Consequently, we now have a 100-plus page document that tries to cover all the loopholes. But state implementation is still there, in addition to long deadlines, and that, thank heavens for democracy, is the fly in the balm.

The new law does put some parameters on the state's contributions to clean air, however. It requires the states to reduce smog by 15 percent in the first six years and 3 percent annually after that until safe health standards are attained. Other mandates of the new Clean Air Act include:

- Controls on smaller pollution emitters, such as printers, dry cleaners, and bakeries, varied by the current smog levels of cities.
- Limits on auto emissions to be tightened by 1996 to 0.25 gpm for hydrocarbons and 0.4 gpm for nitrogen oxides, with graduated increases after that as necessary, feasible, or cost-effective.
- Gasoline sold in the nine smoggiest cities to be reformulated with additives capable of reducing smog-forming and toxic emissions by 15 percent in 1995 and at least 20 percent by 2000.

- Institute a pilot program in California requiring ultra-clean fuels by 2000 for at least 300,000 cars. Twenty-seven other polluted states have options to comply. Fleets of 10 cars or more in those areas must burn ultra-clean fuels by 1998.
- Industrial pollution emitters of 189 airborne toxics are required to install maximum achievable control technology by 2003.
- Coke ovens must be installed to increasingly stringent control devices by 2020.
- Municipal incinerators are subject to similar as industrial emitters restrictions but over a 12-year period. The exception is the 111 dirtiest utility plants (mostly in Appalachia and the Midwest), which have to start cleaning up their act within the next five years. Utility emissions (including mercury and cadmium) are temporarily exempted.
- Halt production of CFCs and halons (used in fire extinguishers) by 2000 (this is an international mandate); HCFCs by 2015; carbon tetrachloride and methyl chloroform solvents by 2000 and 2002 respectively; and mandatory recycling of CFCs in air conditioners and refrigerators starting in 1992.

If the law results in any economic dislocation of workers, a $50 million fund has been set aside for retraining programs during the regular 26 weeks of unemployment insurance.

Automotive emission control: a partial success

One good event has taken place as the result of 35 years of efforts and that is perhaps that the United States has the world's best automotive emissions control. If you've ever been to that 10-million population metropolis of Mexico City, sitting on a 7,800-feet-high plateau, on a hot summer day, you'll come to appreciate our emissions laws. My wife and I had the displeasure of walking along the main avenue in Mexico City one sweltering July day and virtually choked on the clouds of poorly refined petrol pouring out of thousands of exhaust pipes. That evening, we checked out of our hotel and took the late plane out to a more airworthy coastal resort.

Before we, and our Washington godfathers become too smug with our emission-control successes, however, we need to remember that pollution from U.S.-made cars is cumulatively no better than it was in 1970. The reason: there are a million more cars.

And so the fight to reduce automotive emissions of poisonous gases must go on. What a wonderful opportunity for thousands of environmentally conscious automobile shops to go on an emissions-control checking binge! Who wouldn't want to invest a few dollars to ensure that his car operates at peak efficiency and thereby, contributes to the quality of the air we and our families must breathe?

Mind-boggling statistics

To dramatize the importance of support for the Clean Air Act, as well as the entrepreneurial opportunities in tracking down enforcement, take a look at these statistics:

- More than 45 percent of all Americans, but especially those crowded into our biggest cities, are exposed to dangerous air pollution levels.
- Nearly 25 percent of all children in the United States are at risk from breathing our polluted air.
- The United States contains but 5 percent of the earth's population, but consumes 25 percent of the earth's resources.
- By the year 2030, when our children will be in our shoes, the Census Bureau estimates our population will stand at nearly 300 million.

Air pollution is an insidious and often unnoticed form of dirt that we breathe inadvertently and which affects our well-being. Toxic air emissions vary from state-to-state and can be a measure of control or the lack of it. Figure 9-3 shows how states rank in total toxic air emissions, measured in millions of pounds per year.

Rank	State	Emissions
1.	Texas	239.0
2.	Ohio	172.7
3.	Louisiana	138.3
4.	Tennessee	135.0
5.	Virginia	132.4
6.	Michigan	116.4
7.	Indiana	112.9
8.	Illinois	99.2
9.	Alabama	98.3
10.	North Carolina	94.6
11.	Georgia	93.6
12.	New York	89.4
13.	Pennsylvania	87.5
14.	California	82.7
15.	Utah	77.3
16.	South Carolina	64.2
17.	Mississippi	57.3
18.	Arkansas	54.6
19.	Kentucky	51.7
20.	Missouri	50.6
21.	Florida	50.2
22.	Wisconsin	48.7
23.	Minnesota	42.1
24.	New Jersey	42.0
25.	Washington	40.6
26.	Iowa	39.2
27.	Oklahoma	36.4
28.	West Virginia	35.6
29.	Alaska	31.7
30.	Massachusetts	30.1
31.	Connecticut	26.1
32.	Kansas	24.7
33.	Oregon	20.9
34.	Maryland	20.2
35.	Arizona	16.6
36.	Maine	14.6
37.	Nebraska	14.4
38.	New Hampshire	13.0
39.	Colorado	11.0
40.	Delaware	8.0
41.	Rhode Island	5.9
42.	Montana	5.3
43.	Idaho	4.2
44.	New Mexico	3.6
45.	Wyoming	3.2
46.	South Dakota	2.4
47.	Vermont	1.4
48.	Hawaii	1.1
49.	North Dakota	0.9
50.	Nevada	0.7

Fig. 9-3. State ranking of toxic air emissions.

We have the technology and entrepreneurial incentives to help clean our air. All we need to do now is to go out and sell the concept to every American—who breathes. Sounds simple? Hardly. The public pushes for reform. The congress uses its pull and passes a Clean Air Act, complete with amendments (actually afterthoughts once the loopholes were discovered in the original law), sets up an agency to enforce the act (the EPA), and provides enough of the taxpayers' funds to make it work. Voilà! We shall have clean air soon.

Special interests have their say

In this rather naive scenario, something is forgotten. The Big Oil Lobby and the various sycophants who depend on this industry. During the summer of 1990, the House voted a compromise Clean Air Act that calls for a "reformulated gasoline"—conventional gas made less polluting by mixing additives such as ethanol, a corn derivative. The oil industry responded with a million-dollar advertising campaign attacking the mandated mixture as "government gas."

If that were not so serious, it might have been funny, but a sense of humor is not part of the oil lobby's counterattack. The opposition in this case included the producers of ethanol, composed primarily of midwestern farmers who grow the corn. It is easy to see that the production of ethanol, when measured in billions of gallons, makes for some splendid business opportunities for the producers of ethanol. No wonder the oil lobby's public relations people came up with another zinger when they called the compromise Clean Air Act the "Business Opportunity Act of 1990."

Some ethanol is always being sold, but under the Clean Air Act, the 44 cities whose excessive levels of carbon dioxide and smog forces them to make changes in gasoline composition will have to legislate substantial changes. The American Petroleum Institute replied, "A specific recipe for gas is something we're not going to support."

A triumph of sorts

In the many past years during which air pollution has been cussed and discussed, legislation was always aimed at the automobile rather than its fuel. It is to President Bush's credit that he broke with that tradition in the clean air package he proposed in June of 1990. He noted that car manufacturers had already squeezed out almost all they could from tailpipe emissions. Now it was up to the oil companies to clean up their act.

By calling for cleaner fuels to be used in all sorts of cars in the nine major cities with the worst smog problems, administration officials said they would be able to eliminate as much as 50 percent of the remaining smog-forming emissions caused by cars by 1995 and up to 90 percent by the year 2000.

After some seesaw voting in congress that saw methanol additives defeated and ethanol diluted, despite the prospects of an additional $1 billion bushel sale of corn to manufacture ethanol, the newest measure failed in the Senate by 52 to 46. At this point, the White House dropped its opposition to the pro-ethanol amendment.

An interesting event then took place in congress. The legislators saw a joining of two formerly opposite forces, the Farm Belt with environmentalists who favored clean fuels. It was an unlikely alliance, given the usual clash between the two groups over pesticides and land use. Without administration opposition in the person of hatchetman Senator Robert Dole, the farmer-environmentalist alliance steam-rolled the few oil state senators 69 to 30.

Thanks to the united front of environmentalists and agribusiness, the Clean Air Act now has a chance. Automobiles will burn cleaner and factories will spew fewer pollutants into the air. As much as $10 billion in pollution-control devices will be purchased, creating expanded and entirely new companies and consultants. And for the sake of our children, we will be able to breathe easier—as well as they.

10

Education: consumer-conditioning for eco-entrepreneurship

IN THE FINAL ANALYSIS, IN ENVIRONMENTAL ENTREPRENEURING, THE *BIG* factor is the education of the consumer. After all, the customer votes with his dollar at the cash register. In Europe, the "green" movement has gone through a 20-year evolution. From a fringe movement of a band of hippies to a political power recognized by virtually every central government. In the United States, environmental education, consciousness, and responsibility are in the hands of private societies and councils. Today, these associations are growing in numbers and clout and, in the process, are attracting powerful political adherents.

While American consumer publications have been popular for decades, corporations and government agencies are only belatedly making themselves heard. Corporations are awakening to the siren call of "green" marketing, while government agencies are motivated by congressional sponsors who see the vote-getting attraction in championing environmental concerns.

Europe leads the way

Great Britain, after years of neglect, has come to grips with environmental measures and is gradually educating its people to the necessity of preserving the island. A Minister of State for the environment, David

Trippier, was installed and, in one of his first publicized statements, stated: "The impact of huge global environmental problems . . . will mean that increasingly political parties will be judged by the strength of their commitment to preserving the balance of nature while maintaining healthy economic growth."

One of the giants active in Great Britain, Procter & Gamble, added an amen to entrepreneurially attuned environmentalism when it described the process as "incremental improvements in a program that essentially will never be finished." That being the case, a realistic appraisal is that entrepreneurs involved in consumer education have a lifetime job ahead of them. For as long as the earth is crowded with people, so also will there be pollution. Consequently, there will always be a need for education and enterprises that make it possible to continue living here in the style we have become accustomed to.

Germany

In Germany, the "greens" have graduated from a raucous hippie crowd to a respectable political entity. They have won over numerous legislators and have been instrumental in passing many laws that mandate improvements of past and present violations by industry as well as individuals. Entrepreneurially, the German "green" movement has brought about some unplanned results: new marketing techniques, shifts in consumer preferences, changes in packaging and distribution, and even new profit opportunities from which American business can draw some positive and profitable conclusions.

One German spokesman, representing Werner & Mertz Gmbh, a major manufacturer of household cleaning products, is marketing manager Rolf Lehmann. He stated, "Only a small portion of . . . environment-conscious consumers will accept higher prices. Industry has to offer ecologically harmless products on the price level of replaced goods. . . . Environmental concern is not necessarily a matter of political colors," explained Herr Lehmann, "the people who buy these products may be politically conservative, while others are green-oriented. Among people conscious of the environment—about one-half of the German population—are considerable differences. This is not a homogenous group, either from an economic or a political viewpoint."

As long ago as 1978, the West German* government caught the environmental fever. They introduced an environmental labeling plan which they called "Blue Angel." Its purpose was to identify those products already on the market or being launched that were environmentally friendly. It took another seven years for the idea to catch on among con-

*Although the problems of "East" and "West" Germany will now have to be addressed by the newly united country, for purposes of this discussion, the old designations will be used.

sumers. By 1988, however, nearly 3,000 products in more than 50 product categories were registered with the government and allowed to display the "Blue Angel" label.

Plans are underway in France, Sweden, and the Netherlands to adopt a similar scheme. Under the supervision of the European Council, whose edicts are to take effect in 1992 or 1993, an EC-wide labeling plan is projected. On the other side of the world, Japan is also considering a "Blue Angel" identification plan.

Eastern Europe

Educating consumers in Eastern Europe, who are just coming out of a 70-year communist anesthesia of environmental neglect, is being revived with the aid of a strange teacher—the U.S. Environmental Protection Agency. With the help of funds from the State Department's U.S.A.I.D. program, the EPA is developing environmental study centers in Eastern European capitals. Budapest, Hungary, is one of the first and principal recipients of such a center.

The EPA-sponsored centers are virtual on-site laboratories. Pollution of air and water is intense in Budapest as everywhere else in Eastern Europe, where production was always quantum leaps ahead of environmental concerns. Now the dirty tide is changing.

Environmental university departments are expected to lead the way toward a more livable and productive Europe—providing hands-on environmental technologies and acting as magnets for students from all over the world, as well as sponsoring environmental conferences and meetings.

From an entrepreneurial perspective, such educational activities will attract foreign tuition payments, payments for meetings, and grants from the more opulent countries that realize that the polluted air over Hungary could be their own smog cloud tomorrow. Such educational innovations could conceivably generate entrepreneurial activities that spread out from learning centers like pebbles tossed into ponds. It happened in Southern California, creating the Silicon Valley; it happened in Massachusetts, where Route 128 has become Boston's high tech town. It could happen tomorrow in Hungary, Germany, Poland, or even in Russia.

What's happening here at home

In the United States, too, a "Green Seal" movement is afoot to identify products benign to the environment. Procter & Gamble's spokesperson Dr. Deborah Anderson said, "tries to use labeling to help educate consumers. As for advertising, we treat environmental responsibility as another product benefit."

Other American companies, such as the fast-growing Wal-Mart retail chain and the discount clothing chain Syms, have used consumer education as a marketing tool for many years. Syms states in all its ads and displays, "An educated consumer is our best customer." Wal-Mart has already asked manufacturers to make identifiable changes in their products and when they do, publicizes these results in their stores. "We see this as providing information to the consumer, not as a marketing promotion," explained Bill Fields, Wal-Mart executive vice president of merchandising and sales.

In the aisles of many Wal-Mart stores are large stand-up signs dedicated to the chain's environmental concerns and to help consumer education. They state:

> "Our Commitment: Land—Air—Water. Our customers are concerned about the quality of our air, land, and water and want the opportunity to do something positive. Together with our manufacturing partners, we'll provide you with information on products which have been environmentally improved."

Earth Day 1990

What began in 1970 as a counterculture movement—often identified with flower children and hippies—has developed by 1990 into an entrepreneurial event. In the process, Earth Day itself, the 1990 version, has become much more effective and a springboard for innumerable innovations and improvements.

General Motors, Union Carbide, Nike, Seiko, Coors Brewing Co., American Telephone & Telegraph Co., MTV, DuPont, Mobile Oil, the Chemical Manufacturers Association, Esprit, Shaklee, Church & Dwight, Kellogg, and many more smaller companies got on the Earth Day bandwagon. April 22, 1990 seemed almost like some kind of an ecological Christmas.

From an almost subversive movement, Earth Day has become so mainstream that the FBI, which in 1970 had surveillance going during the Washington celebration, set up an environmental exhibit in its lobby on Washington's Pennsylvania Avenue. Earth Day 90 has served to popularize environmentalism and the word *ecology* as few other events have done. As such, it opened the door for entrepreneurs large and small, and even more vitally, for the public to pay attention to environmental problems and their solutions. Much progress is still ahead. This was proven in a Spring 1990 survey conducted by the *Washington Post*, shortly before Earth Day's 20th celebration. More than 1,000 adults were asked whether they believe the environment has become worse since 1970. The results give rise to some consternation—and also encouragement for further action. Seventy-three percent of respondents believed the environment has gotten worse since 1970, while 57 percent regard the environmental state of the earth to be critical. When asked to rank other national problems in order of perceived importance, however, the environment fared poorly. This is the order in which this cross section of Americans ranked national problems:

1. Drugs
2. Poverty
3. Federal deficit
4. Environment

On the environmentalists' agenda, three major aims stand out. They are:

1. Worldwide access to birth control.
2. Eliminating nuclear weapons.
3. End use of chemicals that destroy the earth's atmosphere.

Said Earth Day's executive director, Christina Desser, "This isn't just birds and bunnies, it's a quality-of-life issue."

What does all this attention on the environment mean to the smaller entrepreneur? Several hundred small companies are in existence that make or distribute products considered environmentally friendly. Many more services provide repairs and maintenance that will enhance our environment. Most of all, Earth Day has brought into focus to the general public the problems of our physical world. It makes the public more inclined to use services that are beneficial to the environment and buy products that add no toxicity to the earth, air, or water. Opportunities to aid this cause and enjoy the free enterprise system at the same time are everywhere. This book has attempted to spotlight a few dozen.

Earth Day: exhibits and ads

Unlike in 1970, the second Earth Day in Washington, DC on April 4–8, 1990 was accompanied by all the fanfare of an inauguration. The Earthtech '90 Technology Fair and International Forum was sponsored by the Environmental and Energy Study Institute, a nonprofit, educational, and public policy organization in cooperation with the Congressional Environmental and Energy Study Conference. The latter included 335 U.S. senators and representatives.

Sustainable Development became the theme of this four-day eulogy to the environment, which took place on the Washington Mall in the shadow of the prestigious Smithsonian Air and Space Museum. More than 100 of the country's most influential corporations, organizations, and government agencies exhibited technologies, products, and ideas designed to balance earth needs with profit motives. Satellite hookups even brought France and Japan into the fold. The clear picture demonstrated that pollution recognized no borders and that pollution solutions were indeed global.

An Earthtech 90's supplement in the *Washington Post* reported the need for other motivating factors "such as pricing and tax policy, regulations, and other incentive-based systems are needed to promote the technologies and products compatible with sustainable development."

Co-chairmen of the organizing committee were two senatorial environmentalists, John Heinz and Albert Gore, Jr. "Unless we harness the forces of the marketplace," stated Pennsylvania Senator Heinz, ". . . unless the private sector can help us provide new and sustainable technologies, we may accomplish too little too late."

Senator Gore of Tennessee added, "All nations . . . must work and plan together to ensure that economic advancement and environmental enhancement are symbiotic, not parasitic. We would all benefit if advanced nations, in cooperation with major lending institutions, established centers of training at locations around the world to create a core of environmentally attractive technologies and practices. We should also begin to collaborate . . . with developing nations . . . to plan and implement modern, efficient, diverse energy programs . . . backed by international credit arrangements for energy-efficient and environmentally sustainable development projects."

Where do we stand today?

In the final analysis, corporate and government action on behalf of the environment is often led by the actions or inactions of consumers. In few cases is the government the leader—as it might be in more densely populated and more heavily polluted areas, such as Los Angeles, the Netherlands, Denmark, and Switzerland. Consequently, the "greening" of global trade will likewise be affected by the power of the pocketbook.

Epilogue: in the next century

In the remaining pages of this chapter, I'll touch upon what some cynics would label environmentally utopian dreams. Dreams they certainly are, but with will, we *can* make them come true.

The sky may really be falling

At the beginning of this book, I referred to entrepreneurs and the environment as being the oil and water of our economic existence. Mankind has been able to homogenize the two opposing forces over the past several decades, however. I have shown, hopefully, that it *is* possible to take care of our ecology and still turn a neat profit in enterprise. It is inevitable that the homogenation process will occasionally break down and the oil and water separate or the process is not quite completed or outside elements force a separation. Since it took several millennia to create environmental problems, it stands to reason that quick-fix solutions will be neither instantaneous nor complete.

The almost-shadowy "Greenhouse Affect" is still lurking somewhere in the atmosphere. Somewhere, hundreds of miles above our heads. If we continue the progress made by many of our ecology-conscious entrepreneurs in abating the ravages of conscienceless human abusers—whether industrial, commercial, agricultural, or governmental—then we have a chance of preserving our world for our progeny.

Frankly, it is not likely to happen that quickly. Should Americans become concerned and sophisticated enough to stabilize the destruction of our environment, then there is always the chance that some lesser developed people will upset the ecological applecart.

What will happen in Brazil where the world's largest biomass, the Amazon Forest, is gradually being reduced to stumps? What will happen in the Saharan desert region where there is little effort to stop the growth of arid lands? Each square mile of barren land releases more carbon dioxide into the atmosphere as surely as another 1,000 carbon dioxide-spewing automobiles in unregulated countries do.

Without major interference with the "progress" of such environmental despoiling, this is what could happen: Released carbon dioxide from the earth rises. In the past, much of this gas was absorbed by the phytoplanktons of the sea. The pollution and warming of the seas, however, reduced the phytoplanktons' bulk and, therefore, their absorption capacity. Miles up in the atmosphere, this carbon dioxide helped to form a denser-than-usual layer that kept heat within the earth's orbit. These higher temperatures on earth caused the polar ice caps to melt a little. Over a period of a century, enough polar water ran off into the world's oceans to raise the water table by an estimated 50 feet. Conclusion: heavily populated areas on land masses below 50 feet above sea level were inundated. New York, London, Shanghai, and a thousand others will have to move inland, as they would become uninhabitable.

But then, in the year 2100, we will no longer be here to care. Still . . . if we could stem the "greenhouse effect" in our lifetime, we could secure our world for future generations. Isn't it worth the effort?

Four-day workweek: environmental advantage, entrepreneurial opportunity

A generation ago, unions, business associations, and government offices were all hot about a four-day workweek, which was going to be the be-all panacea of the 1990s. Well, the 1990s are here and we don't have a four-day workweek. In fact, many entrepreneurs, as well as employees, wish that they had a five-day workweek.

Compressed workweeks were to give employees more leisure time, and that would open up vast opportunities for additional recreational facilities, short travel, greater automotive use, increased sale of books and games, and fewer heart attacks, ulcers, and nervous breakdowns.

Somehow, between the dream and the reality, Americans got so busy that many more women have gone back to work, many couples are double breadwinners, and others are working overtime to make ends meet or to afford themselves the endless little luxuries of our inflation-prone society.

The environment, meanwhile, suffers as more and more workers commute in private cars. Los Angeles, the chief polluter in the regional South Coast Air Quality Management District and which wears the dubious distinction of being 1990's dirtiest city, has created sweeping legislation designed to mitigate traffic woes and exhaust pollution.

In Chicago and Washington, DC, and continuing in Manhattan's granite canyons, automotive gridlock has become an endemic disease. Clearly something must be done but a four-day workweek is not among the pragmatic solutions. Instead, industry is experimenting with suburban relocations and flextime to stagger traffic congestion, providing services beyond the workplace that will allow parents to travel less and keep more elastic hours, and even telecommunications through which an employee can work at home and transmit data to the central office. One professor has even advocated adopting a performance-based system of working instead of time-based. Imagine, there would be no magic in the holy cow called 40 hours!

Despite the decade-old predictions of research companies, the U.S. Department of Labor's Bureau of Labor Statistics, the U.S. Chamber of Commerce, and the United Auto Workers, the four-day workweek has not come about. The trend perceived during the early 1980s toward a four-day workweek has slowed noticeably. Flexible starting and quitting times seem to be much more popular now.

The San Francisco-based New Ways to Work vocational resource center spokesperson, Suzanne Smith, stated that "When my group was founded in 1972, it was a romantic notion to dream of 32 hours of work for 40 hours of pay. Today, however, the American family needs 60 to 80 hours of paid employment each week to survive."

Entrepreneurial opportunities

In this problem of "socializing" the American work week, a number of opportunities become apparent. Entrepreneurs can consider such environmentally inspired opportunities and see how they can be developed into profitable businesses that will also benefit the environment. The *green* benefits might not always be obvious, but then the ingenious entrepreneur will have little problem straining environmental rationales out of these opportunities. It is not a cynical approach, but a realistic one. A business that comes across as being environmentally friendly, picks up a public relations plus that translates into more business, while also having a beneficial ecological fallout. Here are a few more pragmatic possibilities:

- More conveniently located day-care centers, even those on the premises of major corporations, industrial parks, or office buildings, will cut down the travel time of parents who must drop junior off prior to work and pick the tyke up afterwards.

- Operating flexible, economical jitneys could eliminate 10 to 20 private cars, having a direct impact on pollution and congestion.
- Establishing compact fitness centers on the premises of factories, large offices, and industrial parks will help keep workers in better shape—physically and mentally—and keep them working more effectively. Such facilities might be subsidized by other nearby industries and companies.
- Medical services, usually subsidized or paid for by companies and institutions that bring equipment and personnel to major work places will have a beneficial influence on absenteeism and efficiency.
- Healthy food service facilities near employment centers are always desirable, especially if these facilities emphasize proper nutritional concerns and environmental benefits.

An eco-entrepreneurial wish list

A more dramatic method of emphasizing some of our most pressing ecological problems are listed in this section. The following "wish list" only touches on four enterprises of national importance—newspaper recycling, commuter registry, tire recycling, and a student tree program—only the tip of the iceberg.

Newspaper recycling

If an entrepreneur started a newspaper recycling business that only collected Sunday papers, he or she would only have to work one day a week—yet collect 5 billion Sunday papers each year. Do you know how many trees it takes to print 5 billion Sunday newspapers? two and a half million trees. Each year, our nationwide recycling efforts would create a forest of 2.5 million trees. If it costs but $5 to plant a tree sapling, the Sunday newspaper recycling business would be worth $12.5 million a year.

Commuter registry

A commuter registry could be a positive way to cut down automobile traffic and pollution. Every workday, millions of men and women in the United States drive to work—by themselves. All these rugged individualists want to come and go when it suits them. But just suppose that we could establish a *nationwide computerized commuter registry*, mandated by congressional law. Each and every commuter now using 500 gallons of gas annually to get to work would mean a nationwide savings of 500 million gallons of gasoline—just by doubling up one million of the current commuters and riding a minimum of two per vehicle. The additional benefit: 10 billion pounds of "greenhouse gas" would be kept out of the atmosphere and our dependency on OPEC oil imports would be lowered drastically. Let's charge just a dollar per registrant. Grossing a million-a-year wouldn't be a bad starter, would it?

Tire retirers

Tire recycling can be a gigantic multifaceted business. One of the major fallouts of our automotive civilization is the accumulation of worn-out tires—250 million in the United States alone each and every year. Let's come up with a cost-effective reuse plan to rid our landscape of this eyesore. How about making them into swing slings for all the kiddie jungle gyms? Or just old-fashioned tree swings with chains fashioned out of tough tire strips? Or ship them to the Orient or Africa to be cut into soles for the world's shoes? They would probably yield two or three pairs of Chinese zoris out of each tire. We've seen old tires used as obstacle courses for athletic trainees, fishing reefs, garden planters, and ship bumpers. There is no raw material cost. Could this be a profitable business?

Student tree program

Starting students on a nationwide tree-planting program could add 40 million trees to our landscape each year. There are about 40 million students enrolled in the nation's public schools. We invest around $3,500 per student each year. A national campaign conducted with the cooperation of the Secretary of Education, Secretary of the Interior, and head of the EPA, could allow each student to be furnished with a tree sapling and a nameplate with the student's name on it. Each student would plant his or her tree, under supervision of teachers and volunteers, and attach the nameplate. What a learning experience and exercise in participatory democracy!

In just three years, at a cost of about $3.50 per student (1/10 of one of each school district's average expenditure per student), 120 million trees could be planted, of which perhaps 100 million would survive and grow. Carbon dioxide emissions would be reduced in the United States by 18 million tons; energy consumption would be reduced by 40 billion kilowatt hours with a commercial value of well over $4 billion. The cost: $420 million; the savings, $4 billion; the national profit: $3.58 billion—less our administrative cost.

The following are more grim ecological statistics that green entrepreneurs could turn to his or her own advantage. One needs only to use his imagination.

- Each year we consume 1.5 billion trees for construction, paper products, furniture, firewood, etc.
- Eighty-nine percent of the contents of our mountainous landfills have not degraded. Only 11 percent of refuse (17 million tons) is currently recycled.

- *The Los Angeles Times* uses 420,000 metric tons of newsprint annually, but 83 percent of it is recycled paper. On the other hand, *The Washington Post*, as of mid-1989, uses no recycled newsprint among its more than 1 billion pounds of paper used each year, but it is studying the situation.
- If each American were to plant seven trees a year, it would only equal the per capita consumption of trees for our various needs.
- If you don't like the taste of old motor oil in your drinking water, consider that more than one-third of the 1 billion gallons of old crankcase oil from our vehicles winds up somewhere in our environment. (Water filters, anyone?)
- If you could collect all plastic soda bottles used during a single year in the United States, you would have 22 billion bottles. These could be profitably recycled into plastic lumber products, fiberfill insulation, and other industrial products.

Profits in plowshares

While this is written, wars are going on in Iraq, Kuwait, and Saudi Arabia, in Liberia, in Kashmir and some other trouble spots. Perhaps tomorrow there will be other areas of unrest. But in the aggregate, the major players at martial maneuvering have decided that beating swords into plowshares is worth a try.

What would happen should peace really come? Can the dream of mankind be realized for the first time in the 10,000 years of our past history? Will we then, as environmentalists hope, turn our attention and wealth toward curing and preserving our globe instead of destroying it? Can millions of entrepreneurs who have since time immemorial depended on supplying increasingly sophisticated weaponry to governments large and small, convert their skills and production to peaceful products? Will we be concerned with saving humanity, as well as the ecology, instead of diabolically destroying it? Will we, finally, change an almost instinctive drive to consume our environment and kill the stranger on the other side of the mountain, nurturing the human potential that lasting peace could bring?

These are a lot of questions, a lot of ideology. The answers are almost hopelessly optimistic goals. And yet, such answers are being discussed, dissected, recorded, and conjectured. It takes little imagination to envision the many areas of opportunities that could develop if our world economies were converted to peace:

- Eradication of diseases
- Elimination of illiteracy
- Production for human consumption instead of destruction

- Leveling out the wealth of nations
- Restructuring our unmanageable cities
- Curing environmental devastation

These are but a few of the broad opportunities that open up many avenues for entrepreneurial achievement and profits. Our barriers certainly are not technological; they are political. Nations must assure their internal security; their borders; their conversion plans to peaceful economies. To change the tide of national attention, convincing economic alternatives need to be established so that armament producers no longer see disarmament as a threat.

These are lofty ideals perhaps, but they have been envisioned since biblical times. It is a fundamental psychological tenet that you cannot alter a person's, or even an animal's, anti-social or destructive behavior unless you provide a positive and acceptable alternative—socially, morally, and even financially. This also applies to entire nations and industries. It has a direct relationship to correcting environmental abuses and opening up new and even better entrepreneurial opportunities.

Myth: military spending increases jobs

The Worldwatch Institute in its brilliant report "State of the World," capsulizes the problem as it applies to the major nations of this earth: "The myth of military-led prosperity has obscured the fact that civilian spending creates significantly more jobs."

A study performed during the past decade by Employment Research Associates (ERA) in Lansing, Michigan underscores that 321 out of 435 U.S. congressional districts lost money on the military incomes within their borders. How? Their citizens, employed in various roller-coaster military production facilities, paid taxes that were actually above the income of the local jurisdiction. The ERA called it a "Pentagontax"!

Statistics from virtually all other countries, including Russia and China, confirm that increased military spending does not increase employment nor improve the lot of its citizens. Civilian spending, on the other hand, creates more jobs than defense industries simply because consumer goods production is less capital-intensive. The experience is similar when one compares employment opportunities in the small business sector with that of the giant industries: small enterprises produce more employment. This is a lesson other, less-developed countries are beginning to learn—even the highly centralized ones like those in the communist orbit.

"Plowshares" conversion goal

As a starter, we cannot assume that all military spending and defense operations will cease. They will only be reduced gradually—providing no opportunity for Iraq-style tempests to upset the international applecart. But then, even a rather small conversion from military to consumer spending can add immeasurable benefits to society.

Take, for example, our major cities' crumbling infrastructures. Whether New York, Paris, Mexico City, or Calcutta, all megalopolitan compounds are suffering from deterioration under the pressure of increasing populations. Cities, like landfills, can absorb only so many people before they overflow into unmanageable disaster.

Currently, a little more than $100 billion in federal funds are turned over to our major cities in grants. While this sounds like a tremendous sum, it is actually 25 percent less than federal program supports a decade earlier (when measured in relative dollar value). A committee of major city mayors suggested that an additional $30 billion, taken from converted military spending, would go a long way toward bringing community services, housing, health care, mass transit, and education in line with current needs.

The upheavals caused by a conversion from a military-intensive economy to a consumer economy are much greater in Eastern countries than they would be in the United States. Hungary is cutting armament spending by 35 percent. Czechoslovakia estimates that 120,000 workers will be affected by the transition. Poland, one of the world's leading arms exporters, is having a hard time changing over to a more environment-friendly, consumer-oriented economy. All these countries are receiving much U.S. assistance and are providing opportunities for hordes of Yankee "carpetbaggers"—this time equipped with computers, cash, and entrepreneurial ideas.

Planning is already underway

One of the brightest examples of positive "plowshares" conversion in an otherwise bleak union picture is the plan of the International Brotherhood of Electrical Workers. In 1989, the IBEW conducted a skills audit among its members. Out of this survey came an alternative conversion plan that could serve as a model for other unions. More than 40 alternative product ideas were produced by the union study group. These included production and employment in pollution control equipment, energy-efficiency technologies, and water management systems.

On a higher level, a number of state and federal groups have been formed to deal with arms-to-peace conversions. These include the National Commission for Economic Conversion and Disarmament, the Jobs-With-Peace organization, and in Southern California, The Center for Economic Conversion, which focuses on the Silicon Valley.

Germany and Great Britain have even more groups and more studies to prepare them for a switch to peaceful products. One of the most productive efforts has been the Thorsson Report, or "Report in Pursuit of Disarmament," issued in Sweden as far back as 1984.

In the United States, a model for conversion programs is the Defense Economic Adjustment Act (H.R.101), introduced repeatedly by New York Congressman Ted Weiss. The bill calls for a Defense Economic Adjustment Council to prepare guidelines for conversion. While this bill also provides for grassroots information dissemination, the Swedes, being smaller and more manageable, have established a Military Adjustment Global Information Center at the Uppsala University.

At this point, we can only hope that we do as the Boy Scouts are admonished to do: Be prepared. The conversion process from military production to peaceful production offers the greatest and most pragmatic hope for an ecologically inspired alternative use of our boundless energies.

Feeding a hungry world

The shame of vast populations in Ethiopia, Somalia, and Eritrea, all on the verge of starvation, while huge American grain reserves are sitting idly in depots and airfields, is a well-documented tragedy. The culprits were, and in many areas still are, political boondoggling and lack of overland transportation.

In agriculture, as in any other entrepreneurial activity, the needs of the individual and those of society appear to occasionally conflict. When that happens, regulations are enacted by governments, as they are in the case of pesticide use or water demands. This then is an institutional, rather than a technological problem. Education and communication will go a long way toward resolving the problem. The social costs of a regulatory, or even dictatorial, approach can be too high to bear. Blending the needs of agricultural entrepreneurs and societal-environmental necessities will take wise government. This is the real challenge in the decade ahead.

Wise governmental decisions, which cannot be separated from a sustainable environment, must consider human nature and the demonstrable experiences of history. Entrepreneurs of the soil and rural societies around the world, must have the motivation to live in an environmentally responsible manner. To do so, all governments, especially in the developing world and the Eastern bloc, which have been emerging from 45 to 70 years of communist oppression, must regard the following three basic facts:

1. When governments subsidize farmers so that food prices for urban populations are kept artificially low, incentives for environmental care is removed from the agricultural entrepreneur.
2. In countries where children are an economic asset on the land, telling them not to have more children is economic deprivation.

3. Farmers who are merely tenants, not owners, will not have the incentive to improve their environment.

The solutions to these problems tread on dangerous ideological grounds. They involve societal problems of a market economy, population control, and property entitlement. In this book, I can only touch on related issues, but they affect agriculture, they affect the environment, and they affect the very survival of our species.

Goals for the planet Earth

On a global scale, we can envision four goals toward environmental hegemony. But first, we must admit, through constant education and information, that the Environment is Boss. The universe in which we live is Numero Uno and we are the custodians of it. Entrepreneurs big and small can become the leaders in this replenishing and preservation process—starting with these four goals:

1. *Money*. We need an international, central, political-proof superfund. The annual budget of the present umbrella organization, the United Nations Environment Program, is $30 million. The aid the United States lavished on Germany during the Marshall Plan years amounted to nearly $130 billion (in terms of 1990 dollars).
2. *Information*. To implement a strong and workable global improvement program, we need to spread the word like missionaries would spread the gospel. Preferably such an information system should be independent of national politics, but it should be understood to be neutral and of benefit to all people. The Red Cross and possibly UNICEF come close to such a purpose. Only through education and information dissemination will environmental problems be identified and solutions made workable.
3. *Integrations*. One thing we don't need is the proliferation of petty politics that stem from a myriad of overlapping organizations, all vying for attention and funds. Multiplication of conflicting efforts to solve common problems is very expensive, both in money and human energy. Take, for example, Africa, center for the world's do-gooders. Currently, it is the laboratory for 82 international organizations and more than 1,700 private ones. One tiny African nation with a population of 8 million natives, during one recent year, was the target of 340 different aid projects. Another large and strategically important African country is the recipient of the largest U.S.A.I.D. program, administered by the U.S. State Department and supported with American tax dollars.

4. *Attitudes*. To achieve all of the previous goals and to stop wasting time and money while human lives and potentials are being dissipated, some radical restructuring of attitudes is called for. This could be as difficult, of course, as proscribing that, from now on, every one must write left-handed or speak Esperanto. Governmental, corporate, and entrepreneurial action, however, is only the cumulative effect of a lot of individuals. A massive international recognition of our environmental problems working with massive entrepreneurial action could change the power of the status quo, of ignorance, of blind nationalism, and monolithic institutions. If Russia and its satellites could change as rapidly and as radically as they did in 1990, then there is hope for a timely rescue of the earth.

Appendix A

The environmental state of the states

THE 50 STATES OF THE UNITED STATES OF AMERICA ARE FAR FROM UNANIMOUS in their attitudes and laws regarding the environment. While all adhere to federal legislation, each has supplementary regulations that have actually been growing as federal agencies withdraw from affairs rightfully thought to be administrated by the states. Not surprisingly, some state laws overlap and duplicate federal laws and the red tape involved is often quite frustrating. Environmental concerns are relatively new, after all having been started for the most part after World War II. It is expected that the problems in environmental administration will be untangled in the years to come, but in the meantime, it can be frustrating. Most states environmental involvement is primarily in these 12 areas:

1. Water resource management.
2. Forestry (though only 18 percent are nonfederal).
3. Dam safety (46 states have such laws).
4. Coastal zone management (50 percent paid by federal contributions).
5. Abandoned hazardous waste dumps.
6. Leaking, underground storage tanks.

7. Indoor air pollution.
8. Other air pollution control facilities.
9. Groundwater protection.
10. Toxic air emissions.
11. Property title transfers when environmental or health standards are involved.
12. Permitting process for air emission, water discharges, and solid and hazardous waste disposal.

The environmental climate

The following is a general evaluation of what an entrepreneur can expect to encounter in each of the 50 states and the District of Columbia:

Alabama Xenophobia reigns in Alabama! Out-of-state dumpers of hazardous materials are no longer allowed to bury their loads in Alabama soil. Even out-of-state refuse haulers who own landfills in the state will receive a surcharge on their waste shipments.

Alaska Local environmentalists keep eagle eyes out for oil drillers who might pollute some of the United States' last pristine ecosystems. Despite state approval, check with those environmental agencies that could prove troublesome at a later stage.

Arizona Controls on sulphur dioxide emissions are being more strictly controlled because of the pollution to the Grand Canyon and other nearby natural treasures. Officials are very skittish about air pollution from local asbestos contamination. Be sure to have soil tested before digging or building.

Arkansas Heavily wooded areas are being preserved by selective cutting, especially in the Ozark and Ouachita forests. Avoid clear-cut logging if you wish to avoid environmentalists' ire.

California Unwieldy population growth, especially in Southern California, has resulted in the nation's stiffest anti-smog provisions. There are limitations on everything from backyard barbecuing and gasoline-powered lawn mowers to gas-burning buses, boats, and drive-through restaurants. Check carefully before you build or buy.

Colorado The conflict between the environment and growing population needs killed the Two Forks Dam project. Water conservation is a serious issue in this state, as is wilderness preservation.

Connecticut Water is the big issue here. The state provides loans and grants to encourage upgrading of sewage treatment facilities. Avoid any plans to dump untreated effluent into Connecticut rivers, lakes, or even the Long Island Sound. *Also see Rhode Island.*

Delaware Water and recreation emphasis in this state place restrictions on location and construction of marine-related facilities. Construction is possible, but you must be sure of your project's environmental impact.

District of Columbia The District is tough on recycling. Businesses must separate waste paper from other trash; residents must separate newspapers and yard clippings from their garbage. It offers entrepreneurs both more obligations as well as opportunities.

Florida Because of this state's low topography and the site of the United States' only coral reefs, sensitive underground aquifers, and heavy immigration, Florida has some stringent environmental laws. Be careful to obtain proper building permits and environmental clearances, especially in the areas of the coral reefs and Everglades.

Georgia State lawmakers have become very sensitive to environmental concerns. One law covers a statewide ban on the use of phosphates (in drinking water, laundry detergents, et. al.); another requires liners in all landfills.

Hawaii This island state is exploring several options for alternate energy—reducing dependency on imported and polluting oil. Among the considerations are steam (by heating underground water sources), solar energy, and wind power. Environmentalists wield strong influence here.

Idaho Home of more than 4 million acres of wilderness in national forests. Timber cutting is a sensitive issue with environmentalists seeking to preserve the nation's prime forests, which they feel is threatened by uncontrolled logging.

Illinois This heavily industrialized state is trying to control growth in two opposite areas—in the north around Lake Michigan, which is suffering from high pollution from the 3-million-plus population of Chicago and its environs and in the south where Illinois' last large parcel of wilderness remains.

Indiana Entrepreneurial development is achievable everywhere within the state except along Lake Michigan, east of Chicago. The Indiana Dunes National Lake Shore is being guarded and expanded, as it is one of America's most heavily used recreational areas.

Iowa Industry is not the problem here, but agriculture. A century of cropland erosion and heavy use of pesticides has endangered not only the productivity of the land but the groundwater supply.

Kansas Away from Kansas City, development is reasonable. Watch out, however, in the area around Kansas City, Kansas. Kansas City, Missouri is across the state border. Landfills are thought to have endangered drinking water supply and solid waste disposal is a problem.

Kentucky Industry around Lake Cumberland, in southern Kentucky, is being restricted from dumping toxic wastes into this heavily used recreational area.

Louisiana The offshore area and the wetlands along the southern coast of this state provide one-third of America's commercial seafood as well as one of the most highly developed gas and oil fields in the world. It is bound to attract federal government restrictions in both areas.

Maine Anybody doing business here in beverages of any kind, be reminded of strong anti-trash laws: no plastic six-pack rings, no single-serving juice boxes, and refundable deposits on all nondairy drink containers from single-serving to gallons. *Also see Rhode Island.*

Maryland The Chesapeake Bay is one of America's most valuable fishing and recreational resources. It is also subject to heavy pollution and recent controls. Agricultural runoffs and construction on, or adjacent to, the Bay is very limited and controlled.

Massachusetts Packagers will find in this state the nation's toughest recycling laws: all packaging must either be reusable or made of 50 percent recyclable materials. By 1996, the law will call for packaging materials to be 100 percent recyclable. *Also see Rhode Island.*

Michigan Because so many rivers exist in this state, all of which are perceived as polluted or endangered, federal laws are sought to control 14 of the natural waterways. This will substantially restrict or inhibit development along these prime properties.

Minnesota A persistent legislature already has, or is in the process of, imposing environmental restrictions on the state's considerable wetlands, marshes, and swamps. Builders who drain more than one acre of wetlands are required to replace them somewhere else in the state.

Mississippi At the moment, attention in this state is away from commercial and residential restrictions and is concentrated on its coastline. Here, NASA is planning to establish a rocket test site which could add nearly 3 million pounds of pollutants to the air every year.

Missouri Estimates are that the state will run out of landfills by 1997 unless current dumpings are curtailed. All local jurisdictions are mandated to come up with plans for their own garbage disposal. A dioxin "spill" in 1983 at Times Beach cost $37 million and caused the relocation of 800 families.

Montana This state's major environmental problem involves two endangered animals, grizzly bears and wolves. Expansion in wilderness areas is restricted. Mining and grazing are major industries, but tourism is also a fast-growing one.

Nebraska Current concerns involve a radioactive materials dump site in north-central Nebraska. Plenty of water appears to be available from dam projects in this primarily agricultural state.

Nevada While thought of as primarily a state of gambling centers and deserts, Nevada has 733,000 acres of federally protected forest lands. These are off limits to building and even motoring. Air pollution is virtually nonexistent.

New Hampshire Strong, stubborn environmental groups make expansion here limited to small, "clean" enterprises. Even Route 101, a major cross-state highway in need of expansion, was restricted. Manchester is, however, one of the best places for living in the entire United States. *Also see Rhode Island.*

New Jersey Rising salinity, acid rain, and surface-water pollution are the results of overcrowding in one of America's densest states. The Clean Water Enforcement Act imposes stiff fines and jail terms on polluters. Antipollution devices will be mandatory expense items when doing business in this state.

New Mexico Studies should be read that examine this state's 2,150-foot-deep radioactive waste dump used by 10 nuclear weapons plants. Environmentalists now fear possible groundwater infiltration. Nitrates and hazardous wastes have already been noted in Albuquerque's water supply.

New York Home of the infamous Love Canal where, because of 21,000 tons of dumped chemicals (courtesy of Occidental Chemical Co.), 900 families were evacuated and $150 million spent in cleanup efforts. Currently, emphasis is focused on the Adirondack Mountains, the largest state-run wilderness area east of the Mississippi, limiting construction of vacation and retirement housing for New York City escapees. Nassau-Suffolk counties on Long Island exhibit heavy pollution and salinity. Expect permitting difficulties. *Also see Rhode Island.*

North Carolina Environmentalists, led by the Sierra Club, are seeking Mobil Oil Co. restrictions of offshore drilling. This area is prone to hurricanes that could damage rigs and cause oil spills. Some surface water scarcity and pollution in major developed areas will make this a factor in building and expansion.

North Dakota Wetlands are protected here by federal leverage—farmers who drain lands to plant crops forsake federal crop subsidies.

Flooding in the Fargo-Moorhead-Grand Forks areas needs to be considered in planning.

Ohio Last year was the year of environmental awakening by this state's legislature. Controlling measures will cover out-of-state landfill use, curbside recycling programs, returnable bottles and cans, and building restrictions in the lakefront areas of Lake Erie. Look for laws on the books as well as those that might be pending.

Oklahoma Discoveries of underground water pollution, especially in the Oklahoma City and Tulsa areas, foresage restrictive measures. Local environmental groups are buying up some vital tall-grass areas to preserve open prairie spaces.

Oregon Always a very environment-conscious state, Oregon has new laws that require companies to change existing manufacturing processes. This will include substituting nontoxic, nonpolluting chemicals with safer alternatives and using industrial processes that curtail pollution.

Pennsylvania Old atomic production problems and the unparalleled success of Pittsburgh in cleaning up one of the country's worst pollution environments, make the Keystone State number one in controlling industrial air and water infractions. In 1990, the Air and Waste Management Association held its national convention in Pittsburgh—proof that things have changed. It's a long way since 1960 when Pittsburgh enacted the nation's first anti-smog law. Still, 58 lakes are badly polluted due to acid rain drifting in from neighboring states. Pollution is indeed a national problem. Federal action is expected next.

Rhode Island Auto pollution is the number one enemy here, and laws have been passed to make any vehicle sold here virtually pollution-proof by 1993. Rhode Island will be joined in this control by Vermont, Massachusetts, New Hampshire, Maine, Connecticut, and New York.

South Carolina Ever since Hurricane Hugo, restrictions on the development of the Atlantic seashore have been implemented, especially for new construction. Chemical pollution of groundwaters in the areas of Anderson, Charleston, Columbia, and Florence will be calling for greater chemical pollution controls.

South Dakota Unique in the United States, environmental concerns here center in the Black Hills where gold-mining operations have been revived. The old free-for-all and wildcatting days of mining are over as antipollution restrictions will be enforced. Air pollution is insignificant in Pierre and even Sioux Falls.

Tennessee Despite atomic power plants in the Chattanooga area, air pollution is insignificant, although some carbon monoxide pollution is noted over Nashville. Ironically, the greatest pollution problem encountered in recent years is the massive balloon ascensions during University of Tennessee football games—a practice that has since been halted.

Texas Automotive pollution is getting foremost attention here. By 1991, one-third of most state vehicles will have to convert to compressed natural gas; by 2000, it will be 90 percent. Oil spills and tornadoes, which occasionally plague Texas, are hardly subject to environmental controls, except perhaps in the building codes. Austin, the capital, is regarded as one of America's best cities in which to start and operate a business.

Utah The feature of particular concern in this state is its 75-mile-long, 1,800-square mile Great Salt Lake. Environmental problems focus on this state treasure. Developers want to dam some of the lake to allow fresh water to accumulate for recreation and development; environmentalists have thus far successfully blocked such a move.

Vermont Strong environmentalist pressures have restricted expansion of the Appalachian National Scenic Trail, despite tacit Environmental Protection Agency approval. Their fear is that ski areas would overexpand, destroying the pristine character of the territory along with natural habitats for black bears. *Also see Rhode Island.*

Virginia Like Maryland, Virginia has joined in protecting its portion of the Chesapeake Bay. Development on or near the Bay is very limited in order to stem further pollution and harm to the considerable fish and sea life along the shores. Air pollution throughout the state is insignificant.

Washington Five hundred-year-old stands of forests are a unique feature in this state and the cause of environmentalists' battles as far as the U.S. Congress. Seattle features an outstanding recycling program, spearheaded by Procter & Gamble Co. The city expects to recycle 60 percent of all its waste by 1998. Voluntary measures are being tried first, which will be followed by mandatory ones if the first don't work.

West Virginia The biggest problem in this state is surface mining. Regulations and restrictions are being beefed-up through state laws.

Wisconsin The state's biggest environmental concern is the quality of Lake Michigan's water, which has caused certain fish to be declared inedible—including salmon and lake trout. Madison, the capital, and one of the best American cities for family living and for doing business, is the birthplace of the Sierra Club (John Muir), a big university, and the state government—all clean industries. Despite acid rain in Duluth, St. Cloud, and Wausau, Wisconsin has one of the United States' best environmental efforts record.

Wyoming A tourist-heavy state because of Yellowstone Park, Wyoming's big environmental concern is whether to allow expansion of winter tourism facilities or not. With only a little over half a million inhabitants, this state still has one of the United States' highest growth rates, usually a harbinger of coming regulations.

Sources of information, guidance, and financing

Each of the 50 states, as well as the District of Columbia, Puerto Rico, and the Virgin Islands, has one or more agencies that offer assistance to eco-entrepreneurs in a variety of ways—literature, general information, information about local laws and regulations, energy-saving inducements, loan programs, and companies prominent in environmental developments. Most of this information is free. Many of these jurisdictions also offer periodic seminars and conferences. Local libraries as well as public-access libraries in many of the listed agencies can also be of inestimable help. What follows is a rundown of state agencies that can help assist eco-entrepreneurs.

Alabama

Alabama Development Office, Research Division, 135 S. Union St., Montgomery, AL 36130. (205)263-0048. Information on natural resources, quality of life, demographics, and the economics.

Department of Conservation and Natural Resources, 64 N. Union St., Montgomery, AL 36130. (205)261-3486.

Department of Environmental Management, Solid Waste Division, 1751 W.L. Dickinson Dr., Montgomery, AL 36130. (205)271-7700.

Alabama Recycling Agency Hotline: 1-800-392-1924.

Alaska

Department of Environmental Conservation, P.O. Box 0, Juneau, AK 99811. (907)465-2600.

Department of Natural Resources, 400 Willoughby, Juneau, AK 99801. (907)465-2400.

Department of Commerce and Economic Development, Division of Investments, Pouch DI, Juneau, AK 99811. (907)465-2510. Fisheries Enhancement and Commercial Fishing Loan Programs.

Arizona

Department of Commerce, Energy Office, 1700 W. Washington St., Phoenix, AZ 85007. (602)542-3633. Information on all facets of establishing a business in Arizona, including environmental regulations. (602)542-5371.

Department of Environmental Quality, 2005 N. Central Ave., Phoenix, AZ 85004. (602)257-2300.

Land Department, 1616 W. Adams St., Phoenix, AZ 85007. (602)255-4621.

Arkansas

Department of Pollution Control and Ecology, 8001 National Dr., P.O. Box 9583, Little Rock, AR 72219. (501)562-7444

Business Information Clearinghouse, 1 State Capitol Mall, Little Rock, AR 72201. (501)682-3358. Includes the Energy Assistance Program.

East Arkansas Business Incubator System, Inc., 5501 Krueger Dr., Jonesboro, AR 72401. (501)935-8365. Marketing, Management, Networking, and technical assistance.

Arkansas Industrial Development Commission, Energy Division, 1 State Capitol Mall, Little Rock, AR 72201. (501)682-1370. Assists businesses with energy-related problem-solving and new product development. Makes recommendations for conserving energy and cutting costs.

University of Arkansas, College of Business Administration, Entrepreneurial Service Center, Fayetteville, AR 72701. (501)575-4151. Assistance with business planning, information and management.

California

Department of Conservation, Division of Recycling, 1025 P St., Rm. 401, Sacramento, CA 95814. (916)323-3743.

The Environmental Affairs Agency, P.O. Box 2815, Sacramento, CA 95812. (916)322-5840.

The Resources Agency, 1416 9th St., Rm. 1311, Sacramento, CA 95814. (916)445-5656

Department of Commerce, Business Development Office, 1121 L St., #600, Sacramento, CA 95814. (916)322-5665. Information assistance. Pollution control financing.

Department of Commerce, Office of Small Business, 1121 L St., #501, Sacramento, CA 95814. (916)445-6545. Energy Reduction Loans; New Product Development; Hazardous Waste Reduction Loans; and Farm Loans.

SAFE - BIDCO, 1014 2nd St., 3rd Fl., Sacramento, CA 95814. (916)442-3321 or (800)343-7233. Firms involved with energy product or project are eligible for loans.

California Pollution Control Financing Authority, Alternative Energy Source Financing Authority, 915 Capitol Mall, Rm. 280, Sacramento, CA 95814. (916)445-9597.

California (Los Angeles) Recycling Hotline: 1-800-RECYCAL.

Colorado

Department of Natural Resources, 1313 Sherman, Rm. 718, Denver, CO 80203. (303)866-3311.

Department of Health, Waste Management Division, 4210 E. 11th Ave., Denver, CO 80220. (303)331-4830.

Office of Energy Conservation, 112 E. 14th Ave., Denver, CO 80203. (303)866-2507.

Small Business Hotline, 1625 Broadway, #1710, Denver, CO 80202. (800)323-7798 (in state), 534-2525 (Denver metro area) or (303)892-3825. Assistance in conjunction with Small Business Development Centers in the state.

Colorado (other than Denver) Recycling Hotline: (800)438-8800.

Connecticut

Council on Environmental Quality, 165 Capitol Ave., Rm. 239, Hartford, CT 06106. (203)566-3510.

Department of Environmental Protection, Solid Waste Division, 122 Washington St., Hartford, CT 06106. (203)566-5847.

Connecticut Development Authority, 217 Washington St., Hartford, CT 06106. (203)522-3730. Financial assistance for pollution control and energy conservation projects.

Connecticut Product Development Corp., 93 Oak St., Hartford, CT 06106. (203)566-2920. Financial assistance to get new products produced, marketed, and sold.

Delaware

Department of Natural Resources and Environmental Control, 89 Kings Hwy., P.O. Box 1401, Dover, DE 19903. (302)736-3869.

Delaware Recycling Hotline: 1-800-CASHCAN.

District of Columbia

Conservation Services Division, 613 G St., NW, Washington, DC 20004. (202)727-4700.

Department of Public Works, Office of Policy and Planning, 2000 14th St., NW, 7th Fl., Washington, DC 20009. (202)939-8115

Recycling Information Hotlines:

EPA Emergency: 1-800-535-0202
EPA Asbestos: 1-800-334-8571 (ext. 6741)
Environmental Defense Fund: 1-800-CALL-EDF

Florida

Department of Environmental Regulation, 2600 Blair Stone Rd., Tallahassee, FL 32399. (904)488-4805.

Department of Natural Resources, Marjory Stoneman Douglas Bldg., Tallahassee, FL 32303. (904)488-1554.

Product Innovation Center, The Progress Center, 1 Progress Blvd., Box 7, Alachua, FL 32615. (904)462-3942. Assistance for smaller businesses: Legal, Business Planning, Financial, Information, Management, Networking, and Technical.

Georgia

Department of Natural Resources, Floyd Towers East, 205 Butler St., Atlanta, GA 30334. (404)656-3530.

Institute of Natural Resources, University of Georgia, Ecology Bldg., Rm. 13, Athens, GA 30602. (404)542-1555.

Georgia Productivity Center, Georgia Institute of Technology, Atlanta, GA 30322. (404)894-6101. Assistance with Hazardous Waste Management.

Hawaii

Department of Health, EPHS Division, P.O. Box 3378, Honolulu, HI 96801. (808)548-6410.

Department of Land and Natural Resources, Box 621, Honolulu, HI 96809. (808)548-6550.

Environmental Center, Water Resource Research Center, University of Hawaii, 2550 Campus Rd., Honolulu, HI 96822. (808)948-7361.

Department of Agriculture, Agriculture Loan Division, P.O. Box 22159, Honolulu, HI 96822. (808)548-7126. Includes the Aquaculture Loan Program.

Idaho

Department of Health and Welfare, 450 W. State St., 3rd Fl., Boise, ID 83720. (208)334-5879.

Department of Lands, State Capitol Bldg., Boise, ID 83720. (208)334-3280.

Department of Water Resources, 1301 N. Orchard, Boise, ID 83720. (208)327-7900.

Illinois

Department of Conservation, Lincoln Tower Plaza, 524 S. 2nd St., Springfield, IL 62706. (217)782-6302.

Department of Energy and Natural Resources, 325 W. Adams St., Rm. 300, Springfield, IL 62704. (217)785-2800.

Illinois Environmental Protection Agency, 200 Churchill Rd., Springfield, IL 62706. (217)782-3397.

Inventor's Council, 53 W. Jackson, #1041, Chicago, IL 60604. (312)939-3329. Networking and Technical assistance.

Illinois Department of Commerce and Community Affairs, Small Business Energy Assistance Program, 620 E. Adams St., Springfield, IL 62701. (217)785-2428.

For the Chicago area: State of Illinois Center, 100 W. Randolph St., #3-400. (312)917-3133.

Financial and technical information; energy saving methods; and grants and engineering for energy-saving installations.

Indiana

Department of Natural Resources, 606 State Office Bldg., Indianapolis, IN 46204. (317)232-4020.

Department of Environmental Management, 105 S. Meridian St., P.O. Box 6015, Indianapolis, IN 46206. (317)232-8603.

Department of Commerce, Division of Energy Policy, 1 No. Capitol, #700, Indianapolis, IN 46204. (317)232-8987. Financial and technical information. Loan subsidies, energy conservation promotions.

Iowa

Department of Agriculture, Land Stewardship, and Division of Soil Conservation, Wallace State Office Bldg., Des Moines, IA 50319. (515)281-5851.

Department of Natural Resources, E. 9th and Grand Ave., Wallace Bldg., Des Moines, IA 50319. (515)281-5145.

Department of Economic Development, 200 E. Grand Ave., Des Moines, IA 50309. (515)281-3036.

Kansas

State Conservation Commission, 109 SW 9th St., Rm. 300, Topeka, KS 66612. (913)296-3600.

State Department of Health and Environment, Landon State Office Bldg., 900 SW Jackson St., Topeka, KS 66612. (913)296-1500.

Wichita State University, College of Business Administration, Centers for Entrepreneurship and Small Business Management, 008 Clinton Hall,

Campus Box 147, Wichita, KS 67208. (316)689-3000. Develop and conduct seminars; supply information.

Kansas Technology Enterprise Corp., 112 S.W. 6th St., #400, Topeka, KS 66603-3957. (913)296-5272. Encourages research and development programs.

Kentucky

Division of Waste Management, 18 Reilly Rd., Frankfort, KY 40601. (502)564-6716.

Environmental Quality Commission, 18 Reilly Rd., Ash Annex, Frankfort, KY 40601. (502)564-2150.

Natural Resources and Environmental Protection Cabinet, Capital Plaza Tower, 5th Fl., Frankfort, KY 40601. (502)564-3350.

University of Kentucky, NASA/University of Kentucky Technology Applications Program, 109 Kinkead Hall, Lexington, KY 40506-0057. (606)257-6322. Information available to the business and industrial user: Energy, chemicals, plastics, metals and pollution control.

Kentucky Development Finance Authority, 2400 Capital Plaza Tower, Frankfort, KY 40601. (502)565-4554. The "Crafts Guaranteed Loan Program" provides loans up to $20,000 to qualified craftspersons.

Louisiana

Department of Environmental Quality, Solid Waste Division, 438 Main St., Baton Rouge, LA 70804. (504)342-1216.

State Office of Conservation, P.O. Box 94275, Capitol Station, Baton Rouge, LA 70804. (504)342-5540.

Maine

Department of Conservation, State House Station, #22, Augusta, ME 04333. (207)289-2211.

Office of Waste Reduction and Recycling, 286 Water St., State House Station 154, Augusta, ME 04333. (207)289-5300.

State Soil and Water Conservation Commission, Deering Bldg., AHMI Complex, Station #28, Augusta, ME 04333. (207)289-2666.

University of Southern Maine, Center for Research and Advanced Study, The New Enterprise Institute, 246 Deering Ave., Portland, ME 04102. (207)780-4420. To implement economic development, which enhances the quality of life of Maine citizens.

Maryland

Department of Natural Resources, 580 Taylor Ave., Annapolis, MD 21401. (301)974-3987.

Department of the Environment, 2500 Broening Hwy., Baltimore, MD 21224. (301)631-3000.

Department of Economic and Employment Development, Energy Financing Administration, 217 E. Redwood St., Baltimore, MD 21202. 1-(800)654-7336 (in state) or (301)333-6985. Loans to businesses seeking to conserve energy, co-generate energy, or provide fuels and other energy sources.

Massachusetts

Department of Environmental Protection, Division of Solid Waste Disposal, 1 Winter St., 4th Fl., Boston, MA 02108. (617)292-5961.

Executive Office of Environmental Affairs, Leverett Saltonstall Bldg., 100 Cambridge St., Boston, MA 02202. (617)727-9800.

Massachusetts Business Development Corp., 1 Liberty Square, Boston, MA 02109. (617)350-8877. Provides, among others, long-term loans for new equipment or energy conversion.

Michigan

Department of Natural Resources, Box 30028, Lansing, MI 48909. (517)373-1220, TDD (517)335-4623.

Water Resources Commission, P.O. Box 30028, Lansing, MI 48909. (517)373-1949.

Environmental Research Institute of Michigan, P.O. Box 8618, Ann Arbor, MI 48107. (313)994-1200. Provides analytical and experimental investigations and technical assistance.

Michigan Biotechnology Institute, P.O. Box 27609, Lansing, MI 48909. (517)355-2277. Facilitates commercialization of agrofood, energy, chemical products, waste treatment, forest products, and pharmaceutical industries.

Michigan Energy and Resource Research Association, 328 Executive Plaza, 1200, 6th St., Detroit, MI 48226. (313)964-5030. Provides information and assistance with project proposals.

Minnesota

Department of Natural Resources, 500 Lafayette Rd., St. Paul, MN 55155. (612)296-6157.

Pollution Control Agency, 520 Lafayette Rd., St. Paul, MN 55155. (612)296-6300.

Minnesota Technical Assistance Program, Box 197 Mayo, University of Minnesota, Minneapolis, MN 55455. (612)625-4949, in state: (800)247-0015. Provides information on hazardous waste management.

Minnesota Recycling Hotline: 1-800-592-9528.

Mississippi

Bureau of Land and Water Resources, Southport Mall, P.O. Box 10631, Jackson, MS 39209. (601)961-5200.

Bureau of Pollution Control, Department of Natural Resources, P.O. Box 10385, Jackson, MS 39209. (601)961-5171.

Institute for Technology Development, 700 N. State St., #500, Jackson, MS 39202. (601)960-3615. Information and technical assistance.

Missouri

Department of Conservation, P.O. Box 180, Jefferson City, MO 65102. (314)751-4115.

Department of Natural Resources, P.O. Box 176, Jefferson City, MO 65102. (314)751-3332.

Montana

Department of Natural Resources and Conservation, 1520 E. 6th Ave., Helena, MT 59620. (406)444-6699. Grant and Loan Section: 444-6774.

Environmental Quality Council, State Capitol, Helena, MT 59620. (406)444-3742.

Department of Health and Environmental Sciences, Cogswell Bldg., Capitol Station, Helena, MT 59620. (406)444-2544.

Nebraska

Department of Environmental Control and Land Quality, State House Station, Box 98922, Lincoln, NE 68509. (402)471-2186.

Nebraska Natural Resources Commission, 301 Centennial Mall S., P.O. Box 94876, Lincoln, NE 68509. (402)471-2081.

Nebraska Technical Assistance Center (NTAC), W 191 Nebraska Hall, University of Nebraska, Lincoln, NE 68588-0535. (402)472-5600, or in state: (800)742-8800. Provides short-term assistance for engineering, technical, and scientific problems facing small businesses.

Nevada

Department of Conservation and Natural Resources, Capitol Complex, Nye Bldg., 201 S. Fall St., Carson City, NV 89710. (702)885-4360.

Office of Community Services, Capitol Complex #116, Carson City, NV 89710. (702)885-4908.

Nevada State Development Corp., 350 South Center, Suite 310, Reno, NV 89501. (702)323-3625. Offers fixed asset loans and second mortgage financing.

New Hampshire

Council on Resources and Development, Office of State Planning, 2½ Beacon St., Concord, NH 03301. (603)271-2155.

Department of Environmental Services, Waste Management Division, 6 Hazen Dr., #8518, Concord, NH 03301. (603)271-3503.

State Conservation Committee, Department of Agriculture, Caller Box 2042, Concord, NH 03302. (603)271-3576.

New Hampshire Industrial Development Authority, Four Park St., #302, Concord, NH 03301. (603)271-2391. Provides financing to construct facilities producing electric energy; water-powered electric generating facilities; or facilities for the collection, purification, storage, or distribution of water for use by the general public.

New Jersey

Department of Environmental Protection, 401 E. State St., CN 402, Trenton, NJ 08625. (609)292-2885.

Commission on Science and Technology, 122 W. State St., CN 832, Trenton, NJ 08625. (609)984-1671 or (609)633-2740. Biotechnology, hazardous and toxic waste management, aquaculture, and plastics recycling guidance for small entrepreneurs.

New Jersey Recycling Hotline: 1-800-492-4242.

New Mexico

Environmental Improvement Division, 1190 Saint Francis Dr., Santa Fe, NM 87503. (505)827-2850.

Health and Environment Department, 1190 Saint Francis Dr., Santa Fe, NM 87503. (505)827-2780.

Technological Innovation Program, Anderson School of Management, University of New Mexico, Albuquerque, NM 87131. (505)277-5934. Provides assistance to innovators and entrepreneurs for the commercialization of technology-based ideas.

New York

Environmental Protection Bureau, State Dept. of Law, 120 Broadway, New York, NY 10271. (212)341-2446.

Department of Environmental Conservation, 50 Wolf Rd., Albany, NY 12205. (518)457-5400. Hazardous Waste Programs, (518)457-4138. Provide assistance to reduce, recover, and recycle hazardous waste as an alternative to disposal.

North Carolina

Department of Human Services, Solid Waste Management Branch, P.O. Box 2091, Raleigh, NC 27602. (919)733-0692.

Department of Natural Resources and Community Development, P.O. Box 27687, Raleigh, NC 27611. (919)733-4984.

Center for Applied Technology, East Carolina University, Greenville, NC 27834. (919)757-6708. Assistance in marketing research, computerization, wellness programs, and environmental issues.

Improving Mountain Living Center, Western Carolina University, Cullowhee, NC 28723. (704)227-7492. Assistance in business planning in the Appalachian region.

Science and Technology Research Center, P.O. Box 12235, Research Triangle Park, NC 27709-2235. (919)549-0671. Provides access to ongoing research in federal laboratories nationwide.

North Dakota

Department of Health, Division of Waste Management, Box 5520, Bismarck, ND 58502. (701)224-2366.

Institute for Ecological Studies, P.O. Box 8278, University Station, University of North Dakota, Grand Forks, ND 58202. (701)777-2851.

Center for Innovation and Business Development, P.O. Box 8103, University Station, University of North Dakota, Grand Forks, ND 58202. (701)777-3132. Provides business and technical support services to entrepreneurs, inventors, and small manufacturers.

Ohio

Department of Natural Resources, Fountain Square, Columbus, OH 43224. (614)265-6886.

Environmental Protection Agency, P.O. Box 1049, 1800 Watermark Dr., Columbus, OH 43266. (614)644-3020.

The Thomas Edison Program, 65 East State St., #200, Columbus, OH 43266-0330. (614)466-3086. To generate new technological ideas, new products and processes, and new companies.

Ohio Recycling Hotline: 1-800-282-6040

Oklahoma

Conservation Commission, 2800 N. Lincoln Blvd., Suite 160, Oklahoma City, OK 73105. (405)521-2384.

Department of Health, P.O. Box 53551, Oklahoma City, OK 73152. (405)271-7159.

Technology Transfer Center, OSU District Office, P.O. Box 1378, Ada, OK 74820. (405)332-4100. Assistance to inventors and evaluation of new technologies.

Oregon

Department of Environmental Quality, 811 SW 6th Ave., Portland, OR 97204. (503)229-5696.

Water Resources Department, 3850 Portland Rd., NE, Salem, OR 97310. (503)378-3739.

Oregon Resource and Technology Development Corp. (ORTDC), One Lincoln Center, #430, 10300 S.W. Greenburg Rd., Portland, OR 97223. (503)246-4844. Provides investment capital for early stage business finance, and applied research and development projects that can lead to commercially viable products.

Pennsylvania

Department of Environmental Resources, Fulton Bldg., 9th Fl., Box 2063, Harrisburg, PA 17120. (717)787-1323.

State Conservation Commission, Department of Environmental Resources, Executive House, P.O. Box 2357, 2nd and Chestnut Sts., Harrisburg, PA 17120. (717)787-5267.

Appalachian Regional Commission Program, Department of Commerce, Office of Enterprise Development, 402 Forum Bldg., Harrisburg, PA 17120. (717)787-4791. Financial and technical assistance to new enterprises to meet the needs in their regions.

Energy Development Authority, Department of Energy, P.O. Box 8040, Harrisburg, PA 17105. (717)783-9981. Funds for research and technology to develop and conserve Pennsylvania's energy sources.

NASA Industrial Applications Center, University of Pittsburgh, 823 William Pitt Union, Pittsburgh, PA 15260. (412)648-7000. Provides assistance in every facet of technology transfer.

Revenue Bond and Mortgage Program, Department of Commerce, Bureau of Economic Assistance, Rm. 405, Forum Bldg., Harrisburg, PA 17120. (717)783-1108. Financial assistance to companies adding pollution control equipment.

Puerto Rico

Small Business Development Centers (SBDCs), University of Puerto Rico, Box 5253, College Station, Mayaguez, PR 00709. (809)834-3590. Provides assistance to present and prospective business owners.

Rhode Island

State Water Resources Board, 265 Melrose St., Providence, RI 02907. (401)277-2217.

Department of Environmental Management, 9 Hayes St., Providence, RI 02908. (401)277-2781 or 277-2774. Information on markets of agriculture, seafood, and aquaculture products; pesticide and animal health.

Rhode Island Recycling Hotline: 1-800-RI-CLEAN.

South Carolina

Department of Health and Environmental Control, J. Marion Sims Bldg., 2600 Bull St., Columbia, SC 29201. (803)734-5000.

Division of Energy, Agriculture, and Natural Resources, 1205 Pendleton St., Suite 333, Columbia, SC 29201. (803)734-1740.

Department of Agriculture, Small Farms Program, P.O. Box 11280, Columbia, SC 29211. (803)734-2200. To assist small farm owners in marketing and production techniques.

South Dakota

Board of Minerals and Environment, Joe Foss Bldg., Pierre, SD 57501. (605)773-3151.

Department of Water and Natural Resources (same as previous)

Governor's Office of Energy Policy, 217½ W. Missouri, Pierre, SD 57501. (605)773-3603.

Office of Rural Development, Agricultural Financial Counseling, and Agricultural Loan Participation Program, 445 E. Capitol, Anderson Bldg., Pierre, SD 57501. (605)773-3375. Assistance for rural small businesses.

Tennessee

Department of Conservation, 701 Broadway, Ellington Agricultural Center, Nashville, TN 37204. (615)360-0103.

Department of Health and Environment, 701 Broadway, 4th Fl., Nashville, TN 37219. (615)741-3424.

Energy, Environment, and Resources Center, University of Tennessee, 327 S. Stadium Hall, Knoxville, TN 37996. (615)974-4251.

The University of Tennessee Center for Industrial Services, 226 Capitol Blvd. Bldg., #401, Nashville, TN 37219-1804. (615)242-2456. Helps firms to deal with hazardous waste problems.

Tennessee Recycling Hotline: 1-800-342-4038.

Texas

Division of Solid Waste Management, 1100 W. 49th St., Austin, TX 78756. (512)458-7271.

State Soil and Water Conservation Board, P.O. Box 658, Temple, TX 76503. (817)773-2250.

Texas Conservation Foundation, P.O. Box 12845, Capitol Station, Austin, TX 78711. (512)463-2196.

Texas Recycling Hotline: 1-800-CLEAN-TX.

Utah

Bureau of Solid and Hazardous Waste, 288 N. 1460 W., Salt Lake City, UT 84116. (801)538-6170.

State Department of Natural Resources, 1636 W. North Temple, Salt Lake City, UT 84116. (801)538-3156.

Utah Innovation Center, 419 Wakara Way, #206, Research Park, Salt Lake City, UT 84108. (801)584-2500. Encourages technical innovation and entrepreneurship.

Vermont

Agency of Natural Resources, 103 Main St., Waterbury, VT 05677. (802)244-7347.

Small Business Resource and Referral Service, Division of Economic Development, Pavillion Bldg., 109 State St., Montpelier, VT 05602. (802)828-3221. Offers assistance to smaller businesses.

Virgin Islands

Small Business Development Center (SBDC), University of the Virgin Islands, P.O. Box 1087, St. Thomas, VI 00801. (809)776-3206. Offers technical assistance in financial analysis, loan application processing, and business management.

Virginia

Council on the Environment, 903 9th St., Office Bldg., Richmond, VA 23219. (804)786-4500.

Department of Conservation and Historic Resources, Division of Parks and Recreation, 203 Governor St., Suite 306, Richmond, VA 23219. (804)786-2121.

Division of Litter Control and Recycling, 101 N. 14th St., James Monroe Bldg., 11th Fl., Richmond, VA 23219. (804)225-2667.

Department of Mines, Minerals and Energy, Energy Division, 2201 W. Broad St., Richmond, VA 23220. (804)357-0330. Assists in promoting increased energy efficiency.

Virginia Recycling Hotline: 1-800-KEEP-ITT.

Washington

Department of Ecology, Olympia, WA 98504. (206)459-6000.

Department of Natural Resources, Public Lands Bldg., Olympia, WA 98504. (206)753-5327.

Small Business Development Center (SBDC), Innovation Assessment Center, 180 Nickerson, #310, Seattle, WA 98109. (206)464-6357. Offers assistance to inventors and innovators of all types. For a small fee, inventions are evaluated confidentially.

Washington Recycling Hotline: 1-800-RECYCLE.

West Virginia

Conservation, Education, and Litter Control, 1900 Kanawah Blvd. E., Bldg. 3, Rm. 732, Charleston, WV 23505. (304)348-3370.

Department of Natural Resources, 1800 Washington St., E., Charleston, WV 25305. (304)348-2754.

West Virginia Industrial and Trade Jobs Development Corp., M 146, State Capitol, Charleston, WV 25305. (304)348-0400. Supplements other financial incentive programs to help create jobs.

Wisconsin

Department of Natural Resources, Box 7921, Madison, WI 53707. (608)266-2621.

Land and Water Resources Bureau, Department of Agriculture, Trade, and Consumer Protection, 801 W. Badger Rd., Madison, WI 53713. (608)267-9788.

Wisconsin Conservation Corps, 20 W. Mifflin, Suite 406, Madison, WI 53703. (608)266-7730.

Innovation Service Center, University of Wisconsin—Whitewater, 402 McCutchan Hall, Whitewater, WI 53190. (414)472-1365. Encourages development among inventors with promising new products and ideas.

Business Energy Fund. A program of the Wisconsin Housing and Economic Development Authority, 1 S. Pinckney St., #500, Madison, WI 53703. (608)266-9991. Offers business loans for energy-related improvements and equipment purchase.

Wyoming

Environmental Quality Department, 122 W. 25th St., 4th Fl., Herschler Bldg., Cheyenne, WY 82002. (307)777-7937.

State Conservation Commission, 2219 Carey Ave., Cheyenne, WY 82002. (307)777-7323.

Appendix B

Environmentally friendly companies

THIS CHAPTER LISTS COMPANIES THAT ARE ACTIVELY INVOLVED WITH saving the environment and can serve as an investment guide as well as a roster of firms with whom environmentally concerned entrepreneurs might want to do business. The following companies were among the leaders in environment-friendly products, or efforts to switch over to such products, and packaging identified by Franklin R & D, a Boston investment firm:

- H.B. Fuller, insisting on environmentally sound construction.
- Apple Computer, recycling as many products as it can and inviting environmental groups to use its facilities.
- Polaroid, like Amoco, a pioneer in waste minimization.
- Nucor, a steel mill that uses new, nonpolluting machinery and handles much recycled metal.
- Wal-Mart Stores, one of the nation's largest retailers, has had an on-going campaign of consumer education and environment-friendly product purchasing.
- Ryder Corp. has made laudable efforts to recycle waste oils and solvents.

The waste processing field

For anyone eyeing environmentally active companies, numerous corporations in the waste processing field are worth looking at. *Thomas Register 1990* listed 85 companies that identified themselves as "enviro" corporations by their names alone. At a recent solid waste disposal symposium, the following 11 companies were named and described in some detail—all involved in some form of *green* entrepreneuring:

ACMAT Corporation

A nationwide insurance carrier, trading as ACSTAR Insurance Company, has one of the largest underwriting capabilities in the industry today. Coverage includes asbestos remediation contractors and other environmental and remediation markets. It is capitalized at $46,660 million.

Allwaste Environmental Services

A $177 million company in asbestos abatement, smokestack cleanup, air moving, and equipment manufacturing for the waste disposal industry. The company is also involved in truck tank cleaning, glass and aluminum recycling, and the design and manufacturing of specialized environmental vehicles. ALWS cleans over 160,000 tank trucks a year, as well as rail tank cars.

The Brand Companies, Inc.

This company focuses on asbestos abatement work and is a leader in the industry. It provides services primarily to industrial clients. Forty-nine percent of its shares are held by Chemical Waste Management. It is capitalized at $358 million.

Browning-Ferris, Inc.

The nation's second-largest solid waste management company, operating about 105 landfills. Currently, they are rapidly expanding recycling and resource-recovery service facilities, including infectious waste collection and disposal—a $100 million business for BFI. The company has more than 1.2 million curbside recycling customers. BFI operates more than 8,000 vehicles.

Calgon Carbon Corporation

Manufactures and markets activated carbon and related products worldwide used for purification, filtration, or concentration of gas and liquid mediums, such as in the purification of drinking water and the remediation of contaminated groundwater. CCC does more than $940 million business a year.

Groundwater Technology, Inc.

Provides a wide range of consulting and remediation services for groundwater and soil contamination problems that typically result from leaking underground storage tanks. Risk assessment and human health environment analysis, site assessment, on-site recovery and restoration, and Superfund remedies are among GTI's services in four countries and 51 states of the union, where offices are maintained. Annual volume of more than $100 million with market capitalization, approximately double the volume.

Schuylkill Holdings/Metals

Recycler of lead products, primarily batteries, at facilities in Baton Rouge, Louisiana and Forest City, Missouri. It is a major hazardous waste processor in this sensitive field.

Sevenson Environmental Services, Inc.

An across-the-board service company to the construction industry that specializes in on-site treatment, containment, excavation, and contaminants removal. It closes out and removes contaminated facilities, such as hazardous landfills like the infamous Love Canal site in Niagara Falls, New York. Operates in the United States as well as Canada. It has a high bonding capacity and grosses between $50 to $60 million annually.

Tetra Technologies, Inc.

A specialty chemical and recycling company that recycles hazardous chemicals and treats wastewater effectively. New plants and increased capacity make this more than $60 million company valuable. Its $166 million market capitalization allows it to invest heavily in research and development.

Waste Management, Inc.

WMI is the world's largest provider of solid and chemical waste collection and disposal services. In the United States, the company has a 10 percent market share and operates more than 100 landfills. Subsidiaries operate in Europe and the Middle East, handle chemical waste disposals, and asbestos abatement. One subsidiary, Wheelabrator Technologies, builds and operates waste-to-energy plants. Net revenue is about $5.5 billion with market capitalization at $18.6 billion.

Wellman, Inc.

The country's largest recycler of waste plastics from polyethylene bottles, fiber waste, and film waste. Landfill linings, fiberfill, carpet fibers, pool and pond linings, and various yarns are all manufactured from such materials. Annual revenue is in excess of $800 million with market capitalization in excess of $1 billion.

Earthtech 90 Technology Fair supporters

The stellar array of exhibitors at the Earthtech 90 Technology Fair, and others who invested many thousands of dollars in advertising in the nation's most prestigious journals or on television, indicates the depth of interest and even commitment on the part of Big Business.

The following is an alphabetical list of companies that participated in the events of Earth Day 1990 and which might be of special interest to investors searching for environmentally friendly companies.

Advanced Environment Recycling Technologies, Inc.
Produces Bioplaste for the construction industry, which is made from waste plastic and waste wood fibers, then recycled into a composite.

Aeration Industries International, Inc.
Develops water cleaning technologies around the globe, from the Chesapeake Bay to the Nile in Egypt and Suyong in Korea.

The Aluminum Association
The country's major aluminum can recycler, cooperating with small businesses, organizations, and schools.

American Electric Power
Developed a "clean coal" technology to allow more efficient utilization of coal under stringent environmental standards.

American Forestry Association
Sponsors a global ReLeaf Program to plant 100 million energy-saving trees in American towns and cities by 1992.

American Gas Association
Developed a new natural gas vehicle and refueling station using natural gas in automobiles and trucks.

American Iron and Steel Institute
Developed mechanisms to recover and recycle steel; promotes steel can collection machines; and uses recycled steel in autos.

American Nuclear Society
Sponsors educational programs that show how nuclear medicine detects diseases, kills bacteria, and helps in food preservation.

American Paper Institute, Inc.
Demonstrates how papers can be sorted and recycled, how to organize a recycling program and manage municipal waste.

Archer Daniels Midland
Producers of ethanol, an environmentally friendlier blend of motor fuel and Polyclean, a cornstarch-based plastic degrader.

Arco (Atlantic-Richfield Company)
Developed a refining process called EC-1, a gasoline designed to help reduce vehicular pollution. They have also planted a million trees.

Asea Brown Boveri

Designing waste processing systems that generate power and help alleviate solid waste disposal problems.

AT&T

Works on three major global issues: telecommunications, CFC replacement, and waste minimization.

Battelle

Produces underground storage tank leak detectors; stabilizers of contaminated soil; and oils from hazardous wastes.

Bechtel Group, Inc.

Energy conservation services (with Sycom Corp.) under the Clean Air Act; clean air technologies; magnetic levitation transportation; solar photovoltaic electricity production.

Boeing Support Services

Airplane manufacturing processes that enhance environmental protection through low emission and noise abatement developments.

The Brooklyn Union Gas Company

Six new technologies that reduce greenhouse effect; solar energy production; the first natural gas heat pump; and carbon dioxide conversion into comestible soda water.

Bureau of Land Management

Developed and administers the Automated Lands and Mineral Records System.

California Energy Company

Features renewable geothermal energy sources and its commercial viability in producing electricity.

The Center for Global Change

Demonstrates low-flow showerheads and energy-saving, compact fluorescent light bulbs, as well as "greenhouse boxes."

Chevron Corporation

Underground storage tank improvements that prevent below-surface leakage, hazardous waste management, and recycling offshore drilling rigs as artificial reefs.

The Conservation Foundation

Demonstration of improved land use planning and natural resource conservation efforts.

Crane Co./Chempump Division

Hermetically sealed, single-unit pump and motor that eliminates possible leakage of fluid chemicals that are ecologically harmful.

Daimler-Benz AG

Advertises Mercedes Benz as, "the best automobiles . . . their safety and environmental compatibility."

Degussa Corporation

Demonstration of an in-house wastewater facility, utilizing hydrogen peroxide (H_2O_2) to destroy cyanide, hypochlorite, formaldehyde, phenols, and in odor removal.

Donaldson Company, Inc.

Diesel exhaust particulate trap oxidizer to clean up diesel exhaust and meet 1991 Environmental Protection Agency Standards and beyond.

DuPont Corporation

Development and implementation of improved processes in plastic recycling, agricultural production, bioremedial soil and water work, new oil and coal technologies, and CFC alternatives.

Dusquesne Light Co.

Demonstration of large-scale recycling of coal combustion products.

Edison Electric Institute

Demonstrated a pollution-free electric vehicle, an improved heat pump, induction cooking tops, and compact fluorescent bulbs.

Environmental Law Institute

An established national research and education institute that formulates and administers environmental controls.

Foster Wheeler Power Systems

Model Commerce Waste-to-Energy facility and power systems.

General Motors

Made a public announcement of GM's "20 Years of Environmental Progress" program and 10-point program of environmental consciousness.

Gilbert Emergency Lighting

Electricity saving LED-type signs that use only seven watts of energy instead of 25 and 47 watts and eliminate failure and maintenance.

Hitachi

Manufactures noise-abating Electronic Sound Attenuation Systems, dust-collecting Moving Electrode Type Electrostatic Precipitors, and sewage and garbage disposal systems.

Inmetco
Reclamation of nickel, chromium, and iron from scrap steel through a thermal process reused in steel alloy production.

IBM
Demonstrates use of computers in tracking and monitoring ecological problems and environmental dangers, from animal protection to the greenhouse effect.

Imperial Chemical Industries
Britain's chemical giant promotes its anti-CFC, ozone-protecting new generation of fluorocarbons in big U.S. advertising.

Komatsu
Japanese manufacturer of construction equipment, lasers, and robots, "designed to meet local needs and global concerns for the environment," according to a *Wall Street Journal* ad.

LUZ International Ltd.
Builders of a 40-foot solar collector that produces 300 megawatts of clean power for a Southern California plant.

Microbial Genetics/Pioneer Hi-Bred International
Bioremediation work on contaminated sites.

Montedison
Italian giant chemical company producing technologically advanced, environmentally sound materials and clean energy.

Murray Corporation
CFC-12 recycling and recovery equipment.

National Coal Association
Demonstrated environmental compatibility of coal production and land use, reclamation and wetlands development.

National Institute of Standards and Technology
Conducts environmentally related research on alternative refrigerants, asbestos, radon, and biomonitoring.

Orange and Rockland Utilities, Inc.
Latest energy-efficient fluorescent lighting technology.

OSRAM Corporation
Energy-efficient, compact Dulux E1 lightbulbs, which originated in Germany, that consume 75 percent less energy and last 13 times longer.

Petro Group Inc.
Rapid-response chemical or oil spill equipment, rolls up flat for easy

storage, self-energizing—called Auto Boom. It is used in emergencies like the Valdez oil tanker disaster.

Philips Lighting

Earth Light SL 18, a compact fluorescent lamp to replace regular 75-watt incandescent bulb at a 75 percent energy saving, thus saving considerable electricity.

Pyromid, Inc.

Outdoor cooking systems that use 75 percent fewer briquettes, needs no air-polluting liquid fire starter, and has a recyclable reflective liner and ash/grease catcher.

Safety-Kleen Corporation

Contaminated waste stream management that protects the environment and conserves resources successfully used all over the United States.

Salomon Brothers

Investment house arranged swap of debt-forgiveness between Sweden, the lender, and Costa Rica for a 210,000-acre tropical forest—the first debt-for-nature exchange—thus preserving millions of environmentally valuable trees and fauna.

Siemens Automotive

German-based auto emission controls help auto engines burn fossil fuels more efficiently using sophisticated sensors and, thus, cutting down considerable air pollution.

Simple Green

Makes biodegradable, nontoxic, concentrated, all-purpose cleaners that do not contaminate sewers and streams.

Smith Barney

Brokerage house that specializes in financing solid waste management, recycling, and resource recovery facilities.

Solar Box Cookers International

Offers a solution for fuel wood shortage and environmental damage by means of an alternative to the current, primitive wood stoves used globally.

Solarex

Photovoltaic power modules for remote water pumping, refrigeration, and lighting using Solarex PV systems.

Southeast Paper Manufacturing Company

A major collector, processor, and recycler of old newspapers, marketing newsprint throughout the southeast United States.

Sukup
Developed the Sukup Bug Beater, which vacuums up insects destructive to vegetable crops, thereby reducing pesticide use.

Sun Company
Manufactures EcoClear Diesel, a methanol-blended M-85, and other reformulated gasolines, that reduce vehicular emissions.

Systech Environmental
Manufactures recycling systems applicable to cement manufacturing, using liquid waste products rather than precious fresh water.

Teledyne Control Applications
Makes the Leak Alarm System for pollutants, which detects and locates even minor seepage on subsea, underground pipelines, and storage tanks.

3M Corporation
Manufactures environment-protection devices and emission controls that ensure 70 percent pollutant reduction in three years and 90 percent in ten years.

Toter International
Makes polyethylene rollout carts for commercial and residential recycling and refuse collection.

Toyota
Has developed a flexible-fuel vehicle that can run on any fuel from gasoline to methanol or any mixture of them.

Union Carbide
Following the Bhopal, India disaster, this company launched a major drive for "environmental excellence," attempting to eliminate carcinogenic pollution to the environment.

Valspar Corporation
McCloskey Varnish Division has developed products with minimal ozone-producing solvents, named Clean Air Formula.

Virginia Power
Spotlights the Chisman Creek Superfund Site, the nation's only fly ash disposal site on the federal cleanup list.

Westinghouse Electric
Builds waste-to-energy systems that convert garbage into electricity, destroying hazardous wastes with nearly total efficiency. Provides site remediation and spill response on a 24-hour emergency basis.

Appendix C

Organizations for the green entrepreneur

OF THE SEVERAL DOZEN ASSOCIATIONS AND ORGANIZATIONS DEVOTED to environmental concerns, I have selected those that can be of some interest to environmentally concerned entrepreneurs. While none of them seem to be directly concerned with environment-for-profit, many will have information, pamphlets, books, newsletters, and even networking opportunities, of benefit to the entrepreneur. Membership in one or several of these organizations can be a good public relations value as well as answer the social and environmental interests of the businessperson. The addresses given and minimum dues or contributions reflect the latest information available as of late 1990.

Acid Rain Information Clearinghouse, 33 S. Washington St., Rochester, NY 14608. Library, conferences, seminars, research.

Alliance to Save Energy, 1725 K St., NW, Washington, DC 20006. Publications and software on energy-efficiency problems and solutions. Supported by organizations and corporations.

American Council for an Energy-Efficient Economy, 1001 Connecticut Ave., NW, Washington, DC 20036. Information to stimulate greater energy efficiency in buildings and appliances. *The Most Energy Efficient Appliances*. Industry supported.

American Forest Council, 1250 Connecticut Ave., NW, Washington, DC 20036. Promotes good forest-management practices. Publishes *American Forest Council* magazine (monthly). Membership available.

American Forestry Association, 1516 P St., NW, Washington, DC 20005. A public information group that promotes "Operation ReLeaf" and publishes *American Forests* magazine, "Resource Hotline." $24 a year.

America The Beautiful Fund, 219 Shoreham, Washington, DC 20005. Awards community groups for local environmental improvement projects; provides seed donations for food growing. Publishes *Better Times* newsletter. $5+ membership.

American Rivers, 801 Pennsylvania Ave., SE, Washington, DC 20003. Works to preserve rivers and streamsides; publishes *American Rivers* newsletter, outfitters that support conservation.

Better World Society, 1100 17th St., NW, Washington, DC 20036. Worldwide public education to divert nations away from military expenditures toward ecological projects. Nonpolitical. $20+.

Bio-Integral Resource Center, P.O. Box 7414, Berkeley, CA 94707. Help on pest-management problems; publishes *IPM Practitioner* and *Common Sense Pest Control*. Membership $45.

Citizens Clearinghouse for Hazardous Wastes, P.O. Box 926, Arlington, VA 22216. Monitors pollution of any type; promotes recycling; networks with other groups. Sixty-title resource list, $2 to $35 per title; dues $15 to $150. Offices in Birmingham, AL; Riverside, CA; Spencerville, OH; Wendel, PA; Grand Prairie, TX; and Christiansburg, VA.

Clean Water Action Project/Clean Water Fund, 317 Pennsylvania Ave., SE, Washington, DC 20005. Clean, safe water through toxic hazard control, water pollution monitoring, protecting natural resources. Publishes *Clean Water Action News* newsletter. $25.

Co-Op America, 2100 M St., NW, Washington, DC 20036. Attempts to solve environmental crisis by changing the way business is done. Maintains Travelinks travel service, insurance, credit union, and publishes *Building Economic Alternatives* magazine. $20 membership.

Council on Economic Priorities, 30 Irving Place, New York, NY 10003. Nonprofit research organization covering corporate responsibility, environmental issues, national security; publishes *Shopping for a Better World* (sold 650,000 copies). New database of 5,000 details corporate, societal and environmental performance of companies. Membership $25+.

Cousteau Society, 930 W. 23rd St., Norfolk, VA 23517. Research, lectures, books, publications, and TV specials aimed at improving and preserving the environment. Over 300,000 members; $20+ membership.

Cultural Survival, 11 Divinity Ave., Cambridge, MA 02138. Harvard-affiliated business consultants that want to import rain products; manage nonprofit nut importing business; publishes *Cultural Survival Quarterly*. $25 membership.

Earth Island Institute, 300 Broadway, San Francisco, CA 95018. Restoration, preservation, and conservation project around the globe; publishes *Earth Island Journal*. $25 membership.

Environmental Action, Inc., 1525 New Hampshire Ave., NW, Washington, DC 20036. Works on environmental issues with congress and environmental projects; publishes *Environmental Action* Magazine. Membership $20.

Environmental Law Institute, 1616 P St. NW, Washington, DC 20036. Law research and education center; gives courses technical assistance and issues many publications and analyses.

Friends of the Earth, 218 D St., SE, Washington, DC 20003. Lobbying, litigation, public information on a variety of environmental issues; publishes *Not Man Apart* newspaper. Merged with Oceanic Society; maintains contacts with offices in 38 countries. 50,000 members. $25 a year.

Greenpeace USA, 1436 U St., NW, Washington, DC 20009. Worldwide, 3-million member environmental action group; publishes *Greenpeace* magazine. Membership $20.

International Alliance for Sustainable Agriculture, University of Minnesota, 1701 University Ave. SE, Room 202, Minneapolis, MN 55414. Promotes economically viable, ecologically sound agriculture worldwide. Publishes "Manna" newsletter. $10+.

International Society for Arboriculture, 303 W. University Ave., Urbana, IL 61801. Advice on tree care and preservation and impact on society. Publishes *Journal of Arboriculture*. $55.

National Audoubon Society, 950 Third Ave., New York, NY 10022. Involved in the conservation of native flora and fauna, land use, resources, and renewable energy. Publishes *Audoubon Magazine*;

travel programs; educational programs. 560,000 members. $35 a year.

National Coalition Against Misuse of Pesticides, 530 Seventh St., SE, Washington, DC 20003. National network for alternative pest management; reducing dependency on toxic chemicals. Wide range of publications from $2 to $25. Membership $20.

National Recycling Coalition, 1101 30th St. NW, Washington, DC 20006. Education, information, lobbying, and annual Congress; assistance to local programs; and publishes *NRC Connection*. Membership $30.

National Wildlife Federation, 1400 16th St., NW, Washington, DC 20036. Six-million-member, society concerned with across-the-board ecological concerns. Publishes *National Wildlife* and *Ranger Rick* (juveniles). Membership $15.

Natural Resources Defense Council, 40 W. 20th St., New York, NY 10011. More than 130,000 membership; $13.5 million budget. Works with EPA, OSHA, Soviet Academy of Sciences on global environmental problems through public education, science, and law. Published over 60 books and position papers. Membership $10.

Pennsylvania Resources Council, 25 W. 3rd St., Media, PA 19063. Sponsors a recycling conference and publishes *Environmental Shopping*, *All About Recycling*, and *PRC News*. Membership $30.

Resources for the Future, 1616 P St. NW, Washington, DC 20036. An environmental research foundation that works on conserving and developing natural resources, pollution problems, waste disposal, toxic abuse. Newsletter *Resources* is free.

Rocky Mountain Institute, 1739 Snowmass Creek Rd., Old Snowmass, CO 81654. Fosters efficient and sustainable use of resources as a means toward global security, community renewal, and resource efficiency. International. Large publication list.

Sierra Club, 730 Polk St., San Francisco, CA 94109. Advocates wilderness conservation and healthy environment concerns more than half a million members. Maintains Legal Defense Fund. Publishes *Sierra* magazine. Membership $33.

Trees for Life, 1103 Jefferson St., Wichita, KS 67203. Sponsors Grow-a-Tree program for children and international food tree growing projects. Publishes *Life Lines*. Membership free.

World Environment Center, 419 Park Ave. South, New York, NY 10016. Serves as a bridge between industry and government to strengthen environmental management, industrial safety, and information exchange and technical expertise. Publishes *The World Environment Handbook*.

World Resources Institute, 1735 New York Ave., NW, Washington, DC 20006. Provides assistance to governments, organizations, and corporations on environmental problems, resource management, and economic growth. Publishes *World Resources Report*.

Worldwatch Institute, 1776 Massachusetts Ave. NW, Washington, DC 20036. Independent research organization that assists government, public and corporate decision-makers with in-depth position papers on environmental issues. Publishes annual 254-page "State of the World" report used in over 1000 universities and translated into many foreign languages. Book included in $25 membership.

Appendix D

Shopping for a better world

THE COUNCIL ON ECONOMIC PRIORITIES IS A NONPROFIT, PUBLIC-INTEREST research organization. Its purpose is to evaluate the policies and practices of U.S. corporations and issues effecting national security, the promotion of corporate social responsibilities, and international peace. The Council is located at 30 Irving Place, New York, NY 10003. Its toll-free national telephone number is 1-800-U-CAN-HELP.

The Council has issued the following, 42-point guide to environmental action. It can be a syllabus for entrepreneurs who want to conduct a business along environmentally friendly lines. In addition, the Council has published the hugely successful *Shopping for a Better World*, a little guide rating the social and environmental performance of more than 150 companies. It can be ordered from the above address for $5.95, postpaid. Lower prices available for bulk orders.

Save energy

1. **Minimize the use of cars and trucks.** Automobiles in the United States spew nearly 1 billion tons of carbon dioxide into the air every year. To help reduce this pollution, use public transportation, ride a bicycle, or walk whenever possible. If

you must drive to work, carpool. If you are considering buying a new car, look for maximum fuel efficiency.

2. **Make sure your home is properly insulated.** The amount of energy that leaks through American windows each year equals the amount of oil that flows through the Alaskan pipeline. Caulk or otherwise seal all leaks. Buy double-glazed or other windows that effectively retain heat inside in winter and outside in summer.
3. **Make sure all appliances you buy are as energy-efficient as possible.** Refer to the American Council for an Energy-Efficient Economy's booklet *The Most Energy-Efficient Appliances*. Refrigerators with the freezer on the side generally use 35 percent more energy than those with one on top of the other. Models with manual defrost use 50 percent less energy than those with automatic defrost.
4. **Conserve energy.** Turn the air conditioner off or down. See if you can survive winter with the thermostat down a few degrees. Always turn off lights when you leave a room.
5. **Try compact fluorescent light bulbs instead of standard incandescents.** Lighting alone uses 25 percent of America's electricity. Compact fluorescents usually last ten times as long as incandescents and are so energy-efficient that they save more than $25 over their lifetime. If you have 15 light fixtures, that's a savings of $375. But the real savings for each bulb is 200 pounds less coal burned every year and less air pollution, acid rain, and global warming.

Save water

6. **Use low-flow showerheads.** These can cut water use up to 75 percent and still deliver a good shower. If you have several people in your house or apartment, you will save over one hundred gallons of hot water a day. If your water is heated with natural gas, those one hundred gallons not used will save you $200 a year in fuel costs. With electrical heating you'll save around $300 every year.
7. **Use water-efficient fixtures.** Flushing your toilet is the single largest use of water in your apartment or house. Using 5 to 7 gallons a flush, your toilet probably uses up some 8,000 or 9,000 gallons a year. If yours has a tank, put in some stones or plastic jugs filled with water to save thousands of gallons a year. Or get a state of the art, low flush toilet. Using 1 or 1.5 gallons a flush, it will save over 7,000 gallons a year.

Avoid toxic pollutants

8. **Cleaning kitchens and bathrooms.** Baking soda with a wet sponge will clean well. Mix it with equal parts of soap. Try 2 teaspoonfuls of borax with 1 teaspoonful of soap and 1 quart of water. If you want, pour the solution into a used spray bottle.
9. **Cleaning glass.** Mix 3 tablespoons of vinegar with one quart of warm water.
10. **Laundry.** Scentless, colorless soap flakes or liquids contain no toxics and are biodegradable. Instead of detergent, try using the same amount of soap with an extra third of a cup of washing soda (sodium bicarbonate) to remove grease and stains. Because the soap may react with residual synthetics in your clothes, first wash all your laundry once in pure washing soda. One cup of vinegar added to the final rinse cycle makes an effective fabric softener.
11. **Choose soap and paper products that have no artificial scent or color.**
12. **Choose organic shaving cream over creams containing ammonia and ethanol.** Or just use a little soap.
13. **Choose organic shampoo or mix your own.** Tom's of Maine and The Body Shop sell organic personal care products. If you want to make your own, there are numerous recipes and blends without toxic chemicals you can try. Here's one: mix one cup liquid Castile soap (made from olive oil and sodium hydroxide) with half a cup of distilled water and one-quarter cup of olive or avocado oil.
14. **Toothpaste.** Mix peppermint extract with baking soda for whiter, brighter, and safer tooth brushing.
15. **Look for carpets made of natural materials such as cotton or wool.** Synthetic carpets and flooring material often contain toxic ingredients.
16. **Use natural and organic alternatives in place of chemical garden and lawn fertilizers.**
17. **Barbecuing causes a great deal of air pollution, which lighter fluids make worse.** If you use a barbecue, avoid self-lighting charcoal. Light coals with paper and wood kindling instead of lighter fluid.
18. **Avoid products with CFCs.** Chlorofluorocarbons (CFCs) are one of the primary causes of damage to the ozone layer. CFCs are used in the manufacture of polystyrene cups, plates, and egg cartons. They are used in refrigerators and air conditioners. Polystyrene is known generically as Styrofoam, a registered

trademark of Dow Chemical. Most dust-off sprays for photographic or electronic equipment and some kinds of insulation contain CFCs. Look for new refrigerators and air conditioners that don't use CFCs. Better yet, use fans rather than an air conditioner. Switch from foam to heavy paper products. Use "*real*" coffee mugs and plates.

19. **Make sure coolants are recycled during repairs on refrigerators and air conditioners in your home and car.** Leaky air conditioners in cars are responsible for a majority of the CFCs released into the atmosphere. An increasing number of auto repair shops are capturing CFCs for reuse.

Reduce waste and recycle

20. **Recycle.** Separate newspapers, bottles, and cans. Recycling saves energy and uses materials otherwise discarded as waste. A soda can made from recycled aluminum uses less than 50 percent of the energy required to make one from raw materials, causes 90 percent less air pollution, and 95 percent less water pollution.

21. **Look for products with the recycled symbol.** Stick to glass, metal or paper, which are all heavily recycled. In 1988, only three percent of the plastic waste produced in the United States was collected for recycling.

Paperboard that may be recycled

Paperboard that is made from recycled material

Samples of coding to identify the type of plastic.

22. **Avoid products with mixed packaging materials.** Combinations of different plastics are nearly impossible to recycle now, yet most frozen and microwavable foods are packaged in three or four kinds of plastic. That should change in the next few years, as companies like Wellman, Inc. and DuPont experiment with methods to separate different kinds of plastic to make them reusable.

23. **Use recycled paper products.** If you don't use recycled materials, you're not really recycling. Without a market demand for recycled materials, recycling will fail or become another subsidized industry. At home and at the workplace, use stationary, toilet paper, and paper for printing made of at least

50 percent recycled fiber. Conservatree is a reliable wholesale recycled paper broker that carries only recycled paper milled with superior pollution controls in place.

24. **Reuse containers and products as much as possible.** Ask friends or neighbors if they need materials you plan to discard. Donate them to organizations to resell or distribute to those in need. For clothes, furniture, and just about anything you can safely clean, buy second hand, at fairs, bazaars, and garage sales.
25. **Buy in bulk.** This reduces excess packaging and saves you money in the long run.
26. **Diapers.** Use cloth ones you wash yourself or use a diaper service. Terry cloth is the best diaper material. Reusable diapers are far cheaper than disposables over the diaper-using life of your baby.
27. **Avoid disposable products.** Buy a pen with a replaceable cartridge or one that is refillable. Use razors with replaceable blades instead of totally disposable ones.
28. **Avoid products with needless, excessive packaging.** Given two equivalent products, choose the one packaged more simply, in a single material.
29. **Bring a string mesh or canvas bag to the store for smaller shopping trips.** If you buy one or two items in a store, put them into your handbag or briefcase instead of taking a plastic bag from the store.

Food

30. **Buy fresh fruits and vegetables.** Whenever possible, shop at green markets, farmers markets, and farm stands. Their produce isn't already wrapped for distribution. And it's usually far more nutritious.
31. **Look and ask for organically grown foods.** They're better for you and growing them doesn't poison our farmland, groundwater, and waterways with harmful chemicals.
32. **Don't be tricked by perfect-looking fruit.** A majority of the chemicals sprayed on them are used for cosmetic purposes.
33. **Buy produce grown domestically and in season.** Chances are, the farther food must be shipped, the more synthetic chemicals for preservation it will contain. Also, food grown in foreign countries is likely to contain chemicals banned in the United States but still sold abroad.

34. **Let stores and manufacturers know how you feel about needless packaging, the importance of recycling, organically grown foods, and all-natural ingredients.** Don't be afraid to ask companies hard questions about their products or operations. Write to company executives and tell them what you like (or don't like) about their policies. The 1990 ***Shopping for a Better World*** has addresses for 168 companies.

Preserve the environment

35. **Don't buy products made from endangered or overexploited species.** Avoid furs, ivory, reptile skin, tortoise shell, and exotic hardwoods.

36. **Start or join a community garden.** In urban areas, gardens bring communities together and bring a bit of nature into the community.

37. **Plant trees.** Trees turn carbon dioxide into oxygen. One growing tree can replace up to 48 pounds of carbon dioxide a year. Trees can prevent erosion and desertification. Well-placed trees around a house can lower cooling needs by 10 to 50 percent.

38. **Support organizations working on causes that you care about.** Become a member or volunteer some of your time. See the annual *Conservation Directory*, $15 from the National Wildlife Federation, 1412 Sixteenth Street, NW, Washington, DC 20036-2266. Or see the *Directory of National Environmental Organizations*, $35 from U.S. Environmental Directories, Box 65156, St. Paul, MN 44165.

39. **Write letters.** Tell your senator or representative how you feel about environmental issues or specific legislation. Write to *Consumer Reports* asking for coverage of the environmental impact of products. Let government officials and business leaders know we need rechargeable batteries, public receptacles for separated garbage, market incentives for alternative power sources, composting of municipal sludge, and much, much more. See the resources box for the National Wildlife Federation's current environmental legislation hotline.

40. **Urge your school(s) to provide environmental education.** We might yet restore a healthy harmony with nature if consideration for the environment in which we live becomes second nature.

41. **Urge your workplace or company to be more environmentally aware.** Is it energy efficient? Does it use recycled paper? Are materials in the waste stream separated?
42. **Sign Earth Day 1990's Green Pledge.** Buy and use products least harmful to the environment and made by companies that promote global environmental responsibility. Write: Earth Day 1990, P.O. Box AA, Stanford, CA 94309.

Appendix E

Transcript of *Profit the Earth*

PROFIT THE EARTH IS A TELEVISION DOCUMENTARY ABOUT USING MARKET forces to protect the environment—and finding business opportunities in the environmental crisis. It presents, in dramatic format, the very essence of this book's theme: entrepreneurism and environmentalism can mix, profitably for both.

If further proof is needed, Hazel Henderson, the futurist who participated in this show, has carried her message of economic transformation to more than 15 countries around the world. What's more, more than 300 U.S. companies have begun environmental programs of their own.

Profit the Earth was produced by the Nebraska ETV Network for the Public Television Outreach Alliance and was originally broadcast April 16, 1990 by Public Broadcasting Service affiliated television stations. What follows is the complete transcript:

Demonstrators: [*chanting in protest to the Exxon Valdez oil spill*] No ocean drilling . . . no oil spilling . . .

Narrator: Protest, confrontation, the courts—these have been the tools of the environmental movement for 20 years. It's been an endless chain of pitched battles—environmentalists versus industry, business versus the government. While they argue, the crisis grows worse.

Industries claim they can't afford to clean up the mess. Environmentalists argue they can't afford *not* to. But the rules may be changing. Environmentalist Dan Dudek is offering business the opportunity to become part of the solution to one problem: acid rain.

Dan Dudek: If I can show them a way to do that—that's not only virtually painless, but out of which many will profit—then I'm afraid that they have no other argument left except, "We don't want to do it."

Narrator: Dan Dudek is a *new* environmentalist. And he's not alone.

In Colorado, entrepreneurs Ken May and Randy Gee are risking everything they own on clean energy—to profit themselves, and to profit the earth.

Ken May: We're farmers. We're farming the sunshine—pioneers just like the same people that settled this country.

Narrator: In Minnesota, 3M is creating a new kind of corporate culture, one that cuts pollution and saves millions of dollars in the process.

Bud Shaver: [*3M Company*] It makes good business sense. We all live on Spaceship Earth here. Whatever I do affects you, what you do affects me. It's the same way in manufacturing.

Narrator: Futurist Hazel Henderson believes it's time to *reshape* our economic system—bring it back in line with nature's ground rules, and nature's limits.

Hazel Henderson: [*speaking to audience*] . . . And I'm not so interested about some other place after I'm dead as I am in making *this* earth a paradise. [*applause*]

Narrator: The people you are about to meet are finding creative new ways to help save our planet by using the tools of our free enterprise system. They're *new* environmentalists, and they're giving the word "profit" a new meaning. [*title*]

Profit the Earth

[*over video, Los Angeles traffic*] Los Angeles, California—home of three-and-a-half million people—depending on jobs, depending on cars, and depending on freeways. But most of all—depending on water.

The city of Los Angeles was planted on a desert. Most of its water comes from hundreds of miles away.

[*over video, Owens Valley*] From places like this—the Owens Valley. This empty basin was once a lake in the middle of a fertile ranching and farming community. But in 1913, a thirsty Los Angeles began draining the Owens River, the source of the lake. Fresh water, flowing 200 miles downhill, gave L.A. a new lease on life. What the city took away turned the Owens Lake to dust.

But L.A. continued to grow. And soon the Owens River wasn't enough. It was only a matter of time before the city looked beyond the Owens Valley—and saw ancient Mono Lake. The lake itself is salty and alkaline, but the streams that feed it are crystal clear—snow melt from the high Sierras. And for 50 years, Los Angeles has been tapping those streams.

Eileen Mandelbaum lives near Mono Lake, where she directs the Mono Lake Committee, an activist group fighting to stop the water diversions.

Eileen Mandelbaum: Well, back in 1941 we would have been sitting under Mono Lake. The lake has constantly shrunk with water diversions. It dropped more than 40 feet in elevation. So as a result, thousands of acres of land have been exposed.

Narrator: As the lake shrinks, strange mineral formations called "tufa" rise from its depths. The water grows saltier and more alkaline, threatening the food supply for 80 species of nesting and migrating birds. And on windy days, alkali dust from the exposed lake bed blows for miles, poisoning the air.

But Los Angeles has a legal right to the water it takes. Ten years of courtroom battles have produced little more than a trickle for Mono Lake. There simply isn't enough water to satisfy the lake and the city.

Zach Willey sees the Mono Lake crisis as an opportunity. He's an economist for the California office of the nonprofit Environmental Defense Fund. And he's working on a plan to save water in a state where many believe much of it is wasted.

Zach Willey: There are 2,100 surface reservoirs in California, all designed to stop water from flowing to the ocean, and using it primarily for agriculture.

Narrator: Agribusiness uses 85 percent of California's water. Cities get the rest. And while city dwellers pay a high price for water, farmers get theirs cheap—and use it freely.

Zach Willey: [*over irrigation of sugar beets*] They use lots of water on crops because it doesn't cost them very much for the water. So there's no economic reason for them to irrigate differently, which costs something.

Narrator: Zach wants to convince farming interests to sell some of their water to Los Angeles for a profit. This will lessen the city's dependence on dying Mono Lake.

Farmers will have less *water*, but more *money* to invest in water-saving technology. And Mono Lake will stop shrinking. Zach's plan is based on self-interest—one of the cornerstones of the new environmentalism. If it works, everybody wins.

[*over video of meeting at L.A. Dept. of Water and Power*] On a September afternoon at the Los Angeles Department of Water and Power, Zach Willey's win/win scenario is taking shape.

Dennis Williams: Clearly there's an issue on how much water it's going to take to protect the environmental resource, and there's a difference of opinion on that, but . . .

Narrator: This is the new environmentalism in action. After a decade of bitter fighting, L.A., and the Mono Lake Committee are finally working together, helping Zach Willey create a relief package for Mono Lake.

Zach Willey: And we've talked to people who grow all kinds of different crops, all the way from vineyards, to cotton, to rice, to alfalfa . . .

Narrator: Zach knows what he's up against. In California it's an unwritten law: *nobody* takes water from agriculture.

[*over video, Zach Willey and Dudley Silvera walking together*] California agribusiness produces 15 billion dollars worth of crops a year. That takes plenty of water. Obviously, a farmer like Dudley Silvera would think twice about selling *his*. But that's just what Zach Willey is asking Silvera to do. Their conversation seems casual, but the stakes are high. Dudley's water rights go back to the Dust Bowl days. Selling a piece of his birthright is *not* a comfortable thought.

Zach Willey: [*to Silvera*] I've had others say, "I'll lease you for three years." And I've had others who've said, "I'll give you the right of first refusal . . ."

Narrator: Zach himself grew up in a ranching community, and he understands a farmer's fears. High pressure tactics will only backfire here.

Zach Willey: . . . I'll do a 35-year lease with a five-year escape hatch, every five years . . .

Narrator: Zach makes it clear that he doesn't want to put anyone out of business. With the profits from the sale of his water, Dudley can invest in high-priced technologies like drip irrigation. Ultimately, he'll use less water on his cotton, without sacrificing an acre of yield. But it's not an easy sell. Farmers haven't forgotten what L.A. did to the Owens Valley.

What's their biggest fear?

Dudley Silvera: Probably setting some kind of a precedent to taking the water away from the farmers and giving it to the city people, and then all of the sudden not being able to stop the runaway train.

Narrator: Some growers, like John Pucheu, would *like* to sell their water, but it's not that simple.

Zach Willey: Well, thank you. [*to John Pucheu, who has just handed him a glass of water*] That's a precious-looking commodity there.

John Pucheu: [*to Zach*] If we can put less water on the land by irrigating at the maximum efficiency, that will contribute to the lessening of our drainage problem.

Narrator: But John lives in a conservative farming community, and neighbors have threatened a lawsuit if any water leaves the district.

Zach Willey: The people that are negotiating with us on Mono Lake are under some social pressure in their own communities. They have people saying things to them in the coffee shop. They're putting up with a certain amount of heat, so in a certain way they're taking a social risk in their communities, and I really respect them for that.

Narrator: But Zach is also taking a risk. Traditional environmentalists accuse him of selling out to agribusiness by offering farmers high prices for their cheap water. To the charge that he's "killing the earth through compromise," Zach replies that compromise is the only way to *save* it.

Zach Willey: I have strong environmental goals. I'm not going to stand by and watch these resources go down the drain, whether global or local. And I'm certainly not going to stand by and watch them go down the drain, whether global or local. And I'm certainly not going to stand by and watch them go down the drain while people insist that the only way that you're going to save them is to litigate or take it away from the other side. I don't believe that. I believe that the way you're going to save them is to deal with the other side.

Narrator: Zach Willey is an optimist. In a state where water is a life and death issue, he believes there's more than enough to go around. If farmers cut their water use by 20 percent, he's convinced there would be enough for wildlife, for cities—and for agriculture. As for Mono Lake . . .

Zach Willey: Even if you told me that it takes five years to do this, if five years of breaking new ground—and at the same time putting together a preservation package for Mono Lake—is what it takes, then that's what we'll do.

Narrator: Zach Willey believes in creating change by appealing to self-interest. But as global resources shrink, self-interest takes on a new meaning—nothing less than the survival of the human species. And Zach says that means earning to live within limits.

Zach Willey: You know, the environmental problem now is very different from when I was a five-year-old. At that time I think there was around nine million people in California. Now there's twenty-eight! And I think we have a lot of work left to do.

"The Acid Test"

Narrator: [*over video, Dan Dudek hailing a taxi cab*] For more than a year, environmentalists Dan Dudek has been waging a campaign in Congress for clean air. Like Zach Willey, he's an economist for the Environmental Defense Fund. Dan's a missionary for the new environmentalism, and one of his converts is the man who wants to be known as the environmental president.

President George Bush: [*making a public statement, June 12, 1989*] Today I am proposing to Congress a new Clean Air Act, and offering a new opportunity. It's time to put our best minds to work, to turn technology and the power of the marketplace to the advantage of the environment.

Dan Dudek: [*appearing before Senate Subcommittee on the Environment, Oct. 4, 1989*] Mr. Chairman, . . . I am Daniel J. Dudek . . .

Narrator: Today, Dan Dudek is testifying for the new clean air bill. He's helped the White House draft a crucial section of that bill, designed to cut acid rain emissions by half within 10 years.

Dan Dudek: . . . We have been consistently attempting to speak with utilities, both privately and publicly. The utilities were largely in a mode of "Hell no, we won't go. We don't want acid rain legislation." . . .

Narrator: Dan wants to put an end to the confrontation. He wants to make industry a partner instead of a foe. That's how the new environmentalism gets results.

[*over video, President Nixon signing Clean Air Act*] In 1970, the approach was different. That's the year Congress passed the *first* Clean Air Act. It made *some* strides, including reductions in deadly lead emissions from automobiles. But it didn't clear the air. Some problems got worse. Acid emissions from coal-burning power plants continued falling to earth as acid rain, killing lakes, trees, wildlife—and endangering human health.

Industry was reluctant to solve the problem, claiming that jobs and profits were jeopardized by the high cost of cleaning up. And government backed off. Frustrated, environmentalists began taking polluters to court to force compliance with the law. The costs have been high in time and money, the results meager.

Dan Dudek: The problem, I think, with all of that is because it's not tapping into changing human behavior—changing how people, how decision makers, how businesses think about the environment.

Narrator: Dan devised a way to give industry a profit incentive to clear the air. It's called emissions trading, and he hopes it will be part of the new Clean Air Act. Here's how it works.

The government sets limits for the amount of sulfur and nitrogen dioxide coal-burning utilities can emit. They are the major causes of acid rain. If one plant reduces its emissions below that minimum, it can sell its leftover pollution allotment, for a profit, to another utility which fails to meet its limit. The company that pays now has an incentive to clean up its act. And periodically, government lowers the emission levels.

Dan Dudek: The overall result is that total pollution is reduced. But it's reduced in a way that gets people to push technology, to innovate, to experiment, and to use the power of a profit motive, of the market, to produce *extra* reductions. Everybody benefits. It's a win/win proposition.

Narrator: But, emissions trading is a controversial idea. Critics accuse Dan of selling out—giving industry a *right* to pollute.

Dan Dudek: We're not talking about rights to pollute here. You don't have to trade. You have the opportunity to do that if you wish, if that can save you money or provide you with other advantages. If not, I resort to being a traditional environmentalist and say, you've got a responsibility to reduce—do it!

Narrator: Dan Dudek is not a government bureaucrat. In fact, he doesn't even live in Washington. Dan and his wife, Christy, own a 200-year-old home in rural Connecticut, far away from the pressures of his environmental mission.

Dan Dudek: [*in backyard of their home*] And one of the reasons we live here is, if we're fortunate enough to have children, I want them to grow up in proximity, in this kind of contact, experiencing the delight of walking down in an evening and seeing a muskrat having its dinner—and the stimulation that that provides.

Narrator: Dan's decision to become an environmental economist goes back to 1968—his Army days in Korea. During garbage detail he would watch Koreans comb through American military trash and reuse what *we* threw away.

Dan Dudek: It was the first large-scale recycling operation I'd ever seen in my life. And when I . . . got out of the Army . . . and I wanted to finish my college training . . . I had to figure out what it was that I was going to do with my life. And that particular incident, that question of this sharp contrast between the United States and Korea—the differences in how people thought of what had value—that led me quite naturally to study economics.

Narrator: Since then, he's taught economics in college, worked on environmental problems in agriculture . . . and on the ozone crisis. Now Dan's drawing on every bit of his experience and knowledge to fight for the new clean air bill in Washington, D.C.

[*Dudek driving his auto*] Today he's heading back there to lobby for the bill . . . The trip has become a weekly grind. When he can, he uses public transportation. It's not always possible. So he's aware of the irony: He's got to share in polluting the earth—to help protect it.

[*over video, Dan entering office building*] In Washington, the mission is to translate the solutions of the new environmentalism into the working language of the law. Dan Dudek does that in partnership with Joe Goffman, EDF's senior attorney. Their days are packed: dozens of phone calls; dialogues with utility executives about emissions trading; visits from the Justice Department about enforcement and penalties for non-compliance; ever-changing business schedules; more phone calls; a meeting to court the support of a consumer group; lunch, on the run; and endless visits to Capitol Hill to lobby for the new clean air bill.

[*video, Dan and Joe in taxi*] Dan and Joe are on their way to meet Dale Curtis. He's legislative aide to Congressman Sherwood Boehlert of New York. The congressman leads the House Working Group on acid rain. They want his support on emissions trading.

Dan and Joe have to make their case to Curtis. Convert the staffer and he may convert his boss. That's the way Washington works.

Dan Dudek: We're partisans—but partisans on behalf of the environment.

Narrator: The congressman's working group on acid rain represents 45 crucial votes.

Joe Goffman: [*EDF attorney*] We're creating a market so that, if you will, cleaned-up air becomes a valuable commodity that somebody can make investments in creating, and actually sell for money.

Narrator: Their pitch: Everybody wins by trading. Pollution levels drop, companies profit, and so does the environment.

Dale Curtis: [*aid to Congressman Boehlert*] That's where you and I get together. . . . How can we take some of these new ideas and, you know, go beyond what we've been able to achieve before. . . ?

Narrator: Dale Curtis seems friendly to the trading idea. Their lobbying effort will be repeated many times before the clean air bill is passed.

[*over video, Dudek again in taxi*] Dan's work extends well beyond America's borders. [*he arrives at office, greets two Russian visitors*] The Soviets want to know more about emissions trading.

Dr. Konstantin Gofman: [*economist, USSR Academy of Sciences*] I have some preliminary talks with my colleagues in Moscow. . .

Narrator: Dr. Konstantin Gofman is a Soviet economist working on environmental policies.

Dr. Gofman: . . . and they support the idea enthusiastically.

Narrator: The Soviets are among the world's most notorious polluters, and acid rain is a global problem. The pollution generated by one country's industries affects the environment of another.

Dan Dudek: If we can get the Soviet Union also experimenting with these methods, then we've built the basis to begin to think about an international, a genuinely global system for managing these pollutants.

Narrator: But Dan's immediate goal has been to get emissions trading into the new Clean Air Act.

Dan Dudek: I want to get this translated into law. When it's in law, I don't think I can wash my hands and walk away from it. That's not the end of it . . . It's like being in a canoe going down the rapids, and your choices are to steer, or not to steer. . . . I don't think I have the choice to get out of the canoe.

I think that in trying to solve environmental problems, I don't think we've tapped a millionth of our potential as a society. And that's what I'm trying to see that we have the opportunity to exploit. I mean, if we have this capability to solve problems, to create new ideas, why should we turn our back on them? Why shouldn't we encourage that to happen? The wider the range of choice, the richer the alternatives, the *lower* the cost. It's very simple.

"Rags to Riches?"

Narrator: [*over video, landfill, Staten Island, New York City*] This is where New York City's garbage goes: Staten Island. Fresh Kills is the nation's largest dump, a mountain of raw garbage rising 250 feet high—22,000 tons of it delivered all day, every day.

Dave Brown: [*supervisor, Fresh Kills Landfill*] We've got it all. We've got everybody from Jersey, Connecticut, Long Island. All these people that work in the city of New York, they generate garbage in the city of New York. And we've got it. We've got it all here in Staten Island. We're not thrilled with it, but we've got it here.

Narrator: Fresh Kills is a towering symbol of America's garbage crisis. Within 10 years it'll be full. Two-thirds of the nation's dumps have already closed. The rest will be filled within the decade. What will we do with our garbage then? Seattle, Washington faces the same dilemma.

Gene Anderson: [*entrepreneur*] What we've tried to do in the past is we've tried to use the landfills as a cure-all for everything. And I don't think we can do that anymore. I think we have to take everything out of the landfills that don't belong in the landfills and find other things to do with it.

Narrator: Gene Anderson believes in the new environmentalism. He's an entrepreneur who knows that caring for the environment can be good for business. Gene runs a reusable cloth diaper service. But, it's had stiff competition.

Gene Anderson: For 30 years, we in the cloth service have been batting our heads, and we've gone maybe from 7 percent to 10 percent of the people that have used our cloth service since the invention of the disposable diaper, they have 85 percent of the business. The diaper industry still has their 10 percent. So it's becoming more and more of a problem. Disposable diapers are the third largest single item that goes into a landfill.

Narrator: Driving to work each day, Gene sees their impact. This Seattle landfill [*in video*] is full—closed for more than a year. Millions of soiled disposables are buried within it. Seattle families use 10 million of them each month. Sealed underground, soiled disposables are a serious health hazard.

Gene Anderson: I got to thinking: What can I do to help stop this? And, of course, I'm not a great scientist or anything else. The only thing I know is diapers.

Narrator: Gene got an idea. Besides cloth, he would start a new venture—recycling disposable diapers. For a fee, the new service would pick up soiled disposables, recycle their materials, and keep them from the landfills.

Gene Anderson: To my knowledge, nobody in the world is doing it other than myself. I think you can clean up the environment and make money off it at the same time. I can see an opportunity here, and I think it's gonna be a big opportunity.

Narrator: There's a strong environmental ethic in the Northwest. Nearly 70 percent of Seattle residents actively recycle aluminum, glass and paper. Why not disposables?

Gene's business is small. He runs it with his brother, wife, and daughter. There are five other employees. The service runs day and night. For more than a year, Gene has been developing his process. His employees launder the soiled disposables. The human waste is separated from the plastic and sent to the city's sludge facility to be made into fertilizer.

As for the plastic, it's sanitized, spin-dried and boxed until a buyer is found for the material. But the market for recycled plastic is glutted. The price has dropped so low that new plastic is cheaper to make. Gene thinks that will change.

Gene Anderson: Plastics, right now, is about at the same place that aluminum was 15 years ago when they said, "Hey, you've got to start recycling aluminum." Now, today, aluminum is worth quite a bit of money to recycle. Fifteen years ago you couldn't give the stuff away.

Narrator: Price and markets aren't the only concern. Disposables are made from two kinds of plastic. So there's a separation problem to be worked out before they can be used to make other durable goods.

[*over video, Gene loading plastic into truck*] Searching for new uses, Gene sends samples of his mixed plastic to other entrepreneurs, and they experiment with it. One makes flower pots from it; another makes wallboard. Like Gene, they are struggling to get established.

Gene Anderson: At least it's a start. And we know that we can find ways of using this plastic. And I think it will come.

Narrator: The absorbent filler in disposables is wood pulp. It's easier to recycle than plastic. Gene takes samples of the cleansed pulp to Weyerhauser Paper for lab testing.

[*voice over Gene greeting Jay Phillips, lab technician*] Weyerhauser uses a million tons of waste fiber a year. They convert it into paper, containers and boxes. To be accepted, Gene's pulp must be free of bacteria and meet other quality standards.

Gene Anderson: [*to Phillips*] What ballpark figure now—what do you think it would be worth if I can get it to your state?

Jay Phillips: [*Weyerhauser lab technician*] Well, it depends upon the end product that it's going to go into, Gene. If the quality of the fiber is high enough, it could be worth in the $300 to $400-a-ton range.

Narrator: To make it profitable, Gene needs to deliver large amounts of pulp. That means collecting more disposables.

[*over video, Gene Anderson talking on phone*] So far, most clients hear about the service by word of mouth. Still, Gene's business has managed to attract 800 customers.

Gene Anderson: Thank God for the people of Seattle. They are worried enough about the ecology. And the greatest remark we get from most of our people is, "Thank God you're doing this. I've been using disposable diapers with a guilty conscience."

Narrator: Customers pay $16 a month for the service. That just about covers Gene's costs. But to make money, Gene needs thousands of customers. But Gene knows he's on the right track. Suddenly he's got competition—big competition. Procter & Gamble, which makes disposables, announced it would begin recycling them, too.

Gene Anderson: I think it's going to come down to where all companies are going to have to start realizing that whatever you manufacture, you're going to have to be responsible for it.

Narrator: Gene's gearing up to expand the business. He's bought land for a new facility and plans to add up to 90 employees. It's a risky undertaking. The process is not yet perfected. The market for plastic is down. Yet, Gene is optimistic that the idea will pay off.

Gene Anderson: If we can work this properly, we're gonna take 50 percent of the third largest single item that goes into landfills and keep them from the landfill. It's going to make this environment better for the next generation. And that's what we've got to work for. People want to do the right thing. If you do the right thing, the money comes.

A Better Sun-Trap

Narrator: [*over video, solar troughs*] Doing the right thing is also on the minds of two Colorado entrepreneurs.

Ken May: [*Industrial Solar Technology*] We're farmers. We're farming the sunshine—pioneers, just like the same people that settled this country.

Randy Gee: [*Industrial Solar Technology*] Solar energy is a preferred option for energy. It doesn't pollute. We don't have to import it. It's a secure source, it's something that we can count on.

Narrator: Randy Gee and his partner Ken May are environmentalists and businessmen, trying to make a living from the sun. Their solar thermal system concentrates sunlight to produce heat. Aluminum collectors track the sun as it moves across the sky, focusing its rays onto tubes filled with water and anti-freeze.

[*over video, detention center facility*] As it circulates, the liquid transfers heat to a tank, providing hot water for the needs of 480 men at this detention center near Denver—and this swimming pool in Aurora. [*large swimming pool facility in video*] The Aurora system was Ken and Randy's first. They put together—and worried over—every nut and bolt.

Ken May: It was really difficult, because it's your baby out there. This is a tough environment—Colorado's strong winds, large hail, feet of snow, and these things just come through it. So we sleep at night now. [*he laughs*]

Narrator: [*over interior, Solar Research Institute*] It's taken years of research to develop a reliable solar film, the crucial reflective material in Ken and Randy's solar collectors. In the world of solar energy, Ken May and Randy Gee have built a better sun-trap. But the world isn't exactly beating a path to their door.

Ken May: The greenhouse effect, the drought—all those things have caught people's imagination. But it hasn't gotten to the point where it affects people's bottom line or influences how they think about the bottom line. That sort of issue's really political.

Narrator: In the late 1970s, the political climate for solar energy was a lot more favorable. During the Arab oil embargo, experimental solar systems were springing up across the country. Tax credits for installing solar systems were generous. Then, the energy prices dropped and the tax credits all but disappeared. One by one, solar businesses fell by the wayside.

Randy Gee: Industries do not just take off overnight. It takes awhile to build a base, to learn from experiences and to make changes. And solar really didn't have that opportunity. Everything happened too fast.

Narrator: It seemed the worst of times for two young engineers to abandon secure government jobs to start a solar business.

Randy Gee: We got into business basically about the time most people were getting out.

Ken May: . . . And for us it was the right time.

Narrator: It was the right time because they had found the right design—a lightweight solar collector they could manufacture cheaply. That was in 1983.

Randy Gee: We worked a lot, but we had the energy at that level. [*Ken laughs*] I think any time you start a company, the energy just—is *there*.

Narrator: [*over video, Ken playing with his children*] For two men with young families, solar spells risk. But Ken and Randy are environmentalists. They want to profit the earth as well as their own pockets.

Ken May: We bet the whole shooting match—the family, the house, the car, the whole bit. And—that's the American way.

Narrator: Ken sees a huge market for his product when he drives through the industrial section of Denver. Bakeries, canneries, grocery stores, meat packing plants—they all use tremendous amounts of hot water. It's heated mostly with natural gas—cheap, but nonrenewable.

Ken May: We're selling our systems purely in commercial terms, in dollar terms, in that we say, "You won't own this system, you won't maintain this system. You won't even know the system is there. But we're going to save you money."

Narrator: But the potential savings hasn't been enough to convince a single plant engineer to "go solar."

Ken May: That's a very common attitude, there's no doubt about it. People think, well, we're going to have to use this energy, solar energy, at some point—but only when the oil runs out. Only when there's no gas. But we're not looking at it that way. We're ready, pretty much ready to go now.

Narrator: On a sunny day in suburban Denver, the energy crisis seems long ago and far away. Though oil spills and air pollution continue, many people think of solar as passé. Randy's wife, Roberta, believes we have a long way to go.

Roberta Gee: A lot of people still don't know what solar energy is. There's a lot to be learned and a lot to be taught. We have a group of people coming over tonight, and some of them I think still don't understand what Randy does for a living.

Narrator: In seven years, Ken and Randy have invested half a million dollars in their solar business. They continue to improve the technology, to make it cheaper and more reliable. In the meantime, they make ends meet by working as solar energy consultants. And as concern for the environment grows, they expect the energy pendulum to swing back in their direction.

Randy Gee: Boy, it'll be great when it does. [*Ken and Randy laugh*] We're ready. I mean, that's the whole idea—that we've done our homework.

Ken May: We've gotta sit back sometimes and just realize what we've accomplished. We left with the idea of producing a better mousetrap—as engineers. And we believe we have a better mousetrap. There's not a system in the world that compares in price or performance to what you're seeing out here. It's part of the solution, potentially anyway. And if we get rich doing it, more people will get involved in doing it, and—we'll all save the world together.

"Keeping Ahead of the Game"

Narrator: [*over video; assembly line, industrial air pollution*] Making money is what industry does best. Profiting the earth is not.

Each year, American industry spews nearly 3 billion pounds of toxic chemicals into the air. Some pollution you can see. Much of it you can't. This 3M plant in St. Paul, Minnesota emits more than 3 million pounds of toxics a year. 3M operates nearly 100 manufacturing plants. It employs a work force of 82,000. The company makes an astounding 60,000 products. 3M alone accounts for 2 percent of all the air pollution American industry emits.

Many corporations complain it costs too much to clean up. 3M realized it might cost *more* in dollars and reputation *not* to. Staying ahead of government regulations has been corporate policy since 1975. It's saving the company millions. And *that* motivates managers like Gene Cross.

Gene Cross: [*manufacturing technology mgr., 3M Printing and Publishing Div., speaking at meeting*] From a corporate standpoint, we've been challenged with two key objectives. One is to reduce by 70 percent our air emissions in 1993. The other is to reduce our total emissions by 90 percent in the year 2,000. That's going to be a real stretch challenge for us from where we are.

Narrator: To make that stretch, 3M researchers work to eliminate toxic emissions from their processes at the very start. It's cheaper and more effective than treating emissions after they've been created.

In 1974, things were different. 3M used the conventional approach—treating pollution at the *end* of the process by installing costly control devices to smokestacks. This cut deeply into company profits. Then the recession hit. 3M Chairman Ray Herzog told all managers to cut their budgets. Environmental chief Joe Ling was caught in a bind between meeting the chairman's goals and the government's regulations.

Joe Ling: [*former director 3M Environmental Engineering*] I said, "Ray, I heard your message, but I'm a little bit different, because all the money I'm spending for the company is trying to meet the government regulations. If I don't spend the money, we're going to violate the regulations. You tell me we gotta meet the regulations, no matter what it is."

"Well," he says, "that's your problem. I thought you were the expert. That's why we hired you. Why don't you solve that problem?"

Narrator: Joe's solution was a new idea for industry: Prevent pollution at its *source* in 3 ways—redesign products and equipment; recycle; or create products that don't pollute in the first place. The new approach would not only *meet* regulations, it could *save* money.

Joe Ling: So our chairman of the board said, "That's okay if you can do that"—and why don't I have somebody do that. He said, "If you know that saves money, why the hell didn't you do that much sooner? You have waited for so long."

I said, "Ray, it's not because I didn't ask about the problem, it's because nobody asked the question."

Narrator: Joe's idea became company policy.

Joe Ling: [*in 1975 3M film, "Pollution Prevention Pays"*] The 3P program is an attempt to encourage involvement by all 3Mers.

Narrator: Pollution prevention *has* paid off. In 15 years, 3Mers have helped the company save nearly half a billion dollars, and cut 450,000 tons of pollution. But 3M still emits 62 million pounds of toxics a year into the air. Now, employees are working to reduce most of *those* emissions by the turn of the century.

Gene Cross: I think people respond to a challenge. If you just do the same thing every day—day-in, day-out—it's very humdrum. This way you get an opportunity to go out and really make a contribution. And it's doing something that's important.

Narrator: Gene Cross's division makes products for the printing and publishing world. Solvents are used to spread materials evenly onto plastic film or paper. In drying, the solvents evaporate and become toxic emissions. The question is how to prevent them.

The answer: Eliminate solvents from the products. Gene's research lab is working on a *solventless process*. It will take up to 10 years and several million dollars to do the job. But 3M policy requires investing both the time and money to make the product environmentally safe—even if it means putting off short-term profit for environmental gain.

Gene Cross: It's one of the areas that we've identified that's important to us in terms of our long-term survival, because you can be much more effective when you plan for change rather than when you react to some external requirement.

Narrator: At 3M's plant, Hutchinson, Minnesota, the problem is different. Videotape *can't* be made without solvents. Each year, the factory pours 15 million pounds of toxics into the air. So instead of removing solvents, plant manager Glen Bloomer decided to recycle them.

Glen Bloomer: [*3M plant manager, surveying construction of recycling system*] We're installing a solvent recovery system. Rather than wasting that money up the smokestack, we're now recycling those solvents and reusing them.

Narrator: The recovery system will reduce the plant's toxic emissions by 80 percent and pay for itself in 4 years.

Glen Bloomer: It's costing us more, but you've got to remember, we're reusing it, and that's going to give us a good return on our investment.

Narrator: Employees at all levels are involved in 3P projects. Carolyn Wolske works on the videotape assembly line checking the quality of plastic cassettes. Faulty parts used to be thrown out. Now they're recycled.

Carolyn Wolske: Through the years we would see this stuff going into the landfills and wonder why couldn't something be done about it. And we didn't realize at the time that we *could* do something about it. So once we organized our team and knew that we could, we went ahead with it.

Narrator: The plant used to pay $800 a month to *dump* the plastic. Now it's shipped to the Resource Recovery Division to create *revenue* for 3M.

Bud Shaver: [*3M Resource Recovery Division*] We found an outlet for it where we're selling it for $1,200 a month. That material is reprocessed, pelletized, and reformed into flower pots and other plastic types of products like that.

Narrator: Saving the landfill fee and generating revenue add $2,000 a month to earnings. The numbers don't seem like much by 3M standards, but it all adds up to huge savings. What's *waste* to 3M can be another company's raw materials.

Bud Shaver: [*in warehouse, to Jack, a 3M employee*] Hey, Jack, how are you this morning? You got some more customers for this? What's this?

Jack: Polyester, Bud. They're going to make some carpeting out of that.

Narrator: In 1989, Resource Recovery converted 3M's industrial by-products into $38 million in income. That was worth 9 cents a share to 3M stockholders, good dividends for the company—and the earth.

Bud Shaver: We all live on Spaceship Earth here. Whatever I do affects you. What you do affects me. Nobody wants anybody to generate emissions and pollute the earth. The thing that has to be done is we have to find better ways to minimize that which we're generating. We have to find better ways to recover it. And this may require a change in our culture, in our habits of what we buy and what we use. But the end result? There are no alternatives.

"Fronting for Mother Earth"

Narrator: [*over video of Kansas State University campus*] Changing our habits—what we use and what we buy—may take us *beyond* economics. On a warm autumn day at the State University in Manhattan, Kansas, environmentalist Hazel Henderson is exploring her favorite subject: What's wrong with economics?

Hazel Henderson: [*speech to Kansas State University audience*] I suppose if you really could calculate the full cost of a can of hair spray driven with a chlorofluorocarbon propellant, it would probably be somewhere around $12,500 a can . . .

Narrator: Hazel believes that the price of a product which destroys the ozone layer should include the cost of the *damage* it does to the earth. *That's* what's wrong with our modern economic system, she says. It *ignores* environmental values.

Hazel Henderson: [*interview*] I feel, in a sense, that I am fronting for Mother Earth. And she is trying to tell us all the time that we don't own this planet. This planet is not private property. . . . We don't *control* this planet. And if we would learn how nature works, she will be abundant for all of us.

[*to Kansas State audience*] . . . And I'm not so interested about some other place after I'm dead as I am in making this earth a paradise. [*audience applause*] Thank you.

Man: [*to Henderson after her Kansas State speech*] I'd like you to meet our Soviet guest. He's most interested that you're going to Moscow.

The Russian: Thank you for your lecture!

Narrator: From Kansas to the Soviet Union, Hazel Henderson finds an eager audience for her message. She's a one-woman think tank, preaching the need for a new economic system based on respect for Planet Earth. Today she'll talk with dozens of people, sign copies of her best-known book. *The Politics of the Solar Age*, and make time for a radio interview.

Interviewer: [*radio interview*] You are now an economic environmentalist—or are you an environmental economist? How would you label yourself?

Henderson: Well, that's really why I prefer to call myself a futurist, because I think that that's a much more open kind of discipline.

Narrator: Hazel believes we're at a crossroads. If we want a future on earth, we must align our economic system with *nature's* ground rules, and *nature's* limits.

Hazel lives on an island, across the bridge from St. Augustine, Florida. It's a world apart—as yet untouched by Florida's population boom. She chooses to live simply. Why own a car if everything you need is within walking distance? And walking, or biking, [*video, Henderson riding bicycle*] means time for thinking.

Hazel Henderson: I call myself an intellectual boutique, [*laughs*] and I'm trying to offer high-quality thinking, and helping other people think through what it is they're doing.

Narrator: Hazel began "thinking through" the connections between our economy and our ecology 25 years ago—in a very different setting.

Hazel Henderson: Well, I was a young mother in New York City, and it was very alienating to be walking around with a voice inside you saying, "The planet is dying." And I don't know why it was that I knew that. All I knew was that the air pollution in New York was similar to that which had caused 4,000 people to die in London in similar sort of conditions. And I had a young baby and I was really, really worried.

And so I began writing letters—you know, the good old American way. I had just become an American citizen—came here from England. And so, with my newfound citizenship, I decided to really test out what it meant to be a citizen.

Narrator: Hazel got a response from New York City Hall—from Mayor Robert Wagner's "Smoke Control Office"—saying there *was* no pollution. It was only *mist* rolling in from the sea.

Hazel Henderson: My apartment was high enough up so that I *knew* that it was coming from the smokestacks of all of these thousands of incinerators all over the city. It was coming from the power company stacks. I mean, I know mist when I see it, and no way it was mist.

Narrator: The "mist" was getting harder to ignore, and a city councilman finally agreed to set up a meeting about the problem.

Hazel Henderson: And, lo and behold, there were about 20 other people there. And we all threw our arms around each other: "Oh, you mean we're not hypochondriacs? There actually is something in the air!" And so we sort of went from there. We organized a group called Citizens for Clean Air.

Narrator: The citizens wasted no time. With the help of a public-spirited New York advertising agency, they published a series of newspaper ads clearly intended to shock.

[*in video, one ad headline above photos of two x-rays of human lungs*]

Introducing
a new disease:
New York Lungs

[*in video, second ad headline beneath photo of man leaning out apartment window, nearly obscured by black smoke billowing from nearby incinerator*]

Tomorrow morning when you get up take a nice deep breath. It will make you feel rotten.

The ads struck a vein of deep public concern. Citizens for Clean Air received sacks of mail and thousands of dollars in donations.

[*over video, Henderson shaking hands with President Lyndon Johnson*] Hazel's New York campaign was part of a national explosion of concern. In 1967, President Johnson signed the Air Quality Act. Hazel was invited to debate economists on environmental issues. She demanded they come up with an agenda for cleaning up the air we breathe.

Hazel Henderson: And they would always say, "Well, she's a nice lady, but she doesn't understand economics. And if only she understood economics, she would know that what she's suggesting is inefficient and uneconomical."

And so, I would go home thoroughly confused, and then I'd realize that I had better start cracking the books, and I had better start finding out what was wrong with economics, because somehow it never occurred to me to deny the evidence of my own senses.

Narrator: The more she read, the more Hazel suspected that outmoded economic beliefs are at the heart of the problem.

Hazel Henderson: And the truth is that all economies are sets of social legislation that sets up the rules. And this weird idea that the economy is some kind of original state of nature is an absolute fallacy!

Narrator: Hazel is convinced that our present economic system is fundamentally at odds with a healthy environment. It supports the burning of polluting fossil fuels, but offers few subsidies for clean solar energy. It provides plenty of cheap water for agribusiness, while lakes and rivers go dry. It ignores the *real* cost of our standard of living. The damage we do the earth as producers and consumers never shows up in the price of a product.

But Hazel says that we, as individuals, have the power to transform our old economy into a new one—if we make use of our leverage as consumers and investors.

Hazel: [*greeting produce clerk in supermarket*] Hi. Do you have any organically-grown produce?

Produce Clerk: No. I don't, ma'am.

Hazel: You don't.

Hazel Henderson: [*voice over as she is seen shopping*] Every time you spend a dollar you can make a vote in the marketplace.

Narrator: To Hazel, that means buying products which don't pollute the air, the land or the water. It means supporting companies which prove they care about the planet—companies which recycle, use renewable energy, and prevent pollution in their own backyards.

Hazel Henderson: [*in the supermarket*] It's kind of the opposite of a boycott. It's more like sort of a *buycott*. And reward the good guys.

Narrator: When consumers become investors, their power increases tremendously. Sharon Glassman is a Florida realtor—Hazel's friend and partner.

Sharon Glassman: . . . And if the consumer knows—"Oh, I have a choice in the stocks I invest in?"—bingo! The light bulb comes on and he starts telling his stockbroker, who may not know . . .

Narrator: Sharon and Hazel are finding practical ways to bring environmental values into the marketplace. Their most ambitious project to date is a venture capital network. It gives start-up funds to innovative new companies devoted to cleaning up the environment.

Hazel Henderson: The sky's the limit in terms of what we can do about all of this. We have all the technology that we need. We know the direction that we need to go. And the question is: How much effort do we want to spend bailing out the old dinosaurs that are crippled, and how much do we want to spend investing in the new babies that need to grow?

Narrator: Hazel Henderson believes we are on the verge of a transformation as profound as the Industrial Revolution. For the past 200 years we've put our energies into dominating the earth—transforming nature's wealth into material goods.

Hazel Henderson: . . .But there are limits, and we've reached those limits now. It's like the human species is coming up to graduation day now, and this planet is our programmed learning environment. Either we're going to make it here, and learn the rules, or we're not going to make it at all.

Narrator: The new environmentalism is dedicated to the creation of a global economy in harmony with the natural world. One in which renewable energy will be harvested fully for the enrichment of our planet. One in which it will simply cost too much to consume endlessly and waste what we consume. One which will truly Profit the Earth.

"Updates"

Narrator: Zach Willey is close to signing his first water contract with California growers—the beginning of a relief package for Mono Lake.

The emissions trading principles devised by Dan Dudek are part of the new Clean Air Act.

Gene Anderson moved into his new plant, and has more than doubled the number of customers using his disposable diaper service.

Ken May and Randy Gee recently sold another solar heating system—their largest yet—to a detention center in California.

3M's Pollution Prevention idea is catching on.

Glossary

acid deposition A complex chemical and atmospheric phenomenon that occurs when emissions of sulfur and nitrogen compounds and other substances are transformed by chemical processes in the atmosphere, often far from the original sources, and then deposited on earth in either a wet or dry fog. The wet forms are "**acid rain**" and can fall as rain, snow, or fog. The dry forms are acidic gases or particulates.

agglutination The process of uniting solid particles coated with a thin layer of adhesive material or of arresting solid particles by impact on a surface coated with adhesive.

Asbestos Hazard Emergency Response Act (AHERA) A 1986 federal law requiring local education agencies to identify asbestos hazards and develop abatement plans.

air pollutant Any substance in the air that could, if in high enough concentration, harm people, other animals, vegetation, or materials. Pollutants can include almost any natural or artificial composition of matter capable of being airborne. They might be in the form of solid particles, liquid droplets, gases, or a combination of these forms. Generally, they fall into two main groups: (1) those emitted directly from identifiable sources; and (2) those produced in the air by interaction between two or more primary pollutants, or by reaction with normal atmospheric constituents, with or without provocation. Except for pollen, fog, and dust, which are of natural origin, about 100 contaminants have been identified and fall into the following categories: solids, sulfur compounds, volatile organic chemicals, nitrogen compounds, oxygen compounds, halogen compounds, radioactive compounds, and odors.

asbestos A material that can be any of several minerals that readily separate into long, flexible fibers suitable for use as noncombustible, nonconducting, or chemically resistant material. The material can pollute air or water and cause cancer or asbestosis when inhaled. The EPA has banned or severely restricted its use in manufacturing and construction.

attenuation The process by which a compound is reduced in concentration over time through adsorption, degradation, dilution, and transformation.

baghouse filter A large, fabric bag, usually made of glass fibers, used to eliminate intermediate and large particles (greater than 20 microns in diameter). It operates similar to an electric vacuum cleaner bag, passing the air and smaller particulate matter while trapping the larger particles.

benzene (C_6H_6) A potent, hazardous gas created by smoking and by cars idling in attached garages. EPA tests have shown that benzene levels are 50 percent higher in homes of smokers. Benzene is dangerous especially because it is a colorless, aromatic but highly toxic and flammable liquid that evaporates easily.

biotechnology Techniques that use living organisms or parts of organisms to produce a variety of products, from medicines to industrial enzymes, to improve plants or animals, or to develop microorganisms for specific uses such as removing toxics from bodies of water or as pesticides.

carbon dioxide (CO_2) A heavy, colorless gas responsible for about 50 percent of the growing gas blanket that envelops the earth and causes a "greenhouse effect." Every year, 6 billion tons of it are added to the atmosphere—25 percent of it from the United States alone. It comes from burning fossil fuels such as coal, oil, and natural gas, and from burning forests.

carbon monoxide (CO) A toxic gas that can escape from clogged or leaking furnaces, chimneys, gas stoves, and car exhaust fumes. High levels can kill; chronic exposure to low doses can lead to gastrointestinal disturbances, reduced stamina and coordination, drowsiness, and headache.

characteristic Any of one of four categories used in defining hazardous waste; ignitability, corrosivity, reactivity, and toxicity.

chlorofluorocarbons (CFCs) A man-made chemical that is the propellant in aerosol cans and mobile air conditioning units. It rises to destroy the earth's ozone layer and is thought to be responsible for 15 to 20 percent of the global warming.

chloroform A liquid derivative of chlorine that can cause unconsciousness in substantial doses. Continued exposure can be carcinogenic. It is a by-product of tap water, especially hot water.

Clean Air Act Authorizes the EPA to list various hazardous air pollutants. New regulations are to include asbestos, beryllium, vinyl chloride, benzene, arsenic, radionuclides, mercury, and coke oven emissions. The Clean Air Act also sets certain emission standards for many types of air emission sources, including standards for stationary sources, motor vehicles, RCRA-regulated incinerators, and industrial boilers or furnaces.

Clean Water Act Lists substances to be regulated by effluent limitations of more than 21 primary industries. The Clean Water Act substances are incorporated into both RCRA and CERCLA. In addition, the Clean Water Act regulates discharges from publicly owned treatment works to surface waters, and indirect discharges to municipal wastewater treatment systems. Some hazardous wastewaters that would generally be considered RCRA-regulated wastes are covered under the Clean Water Act because of the use of treatment tanks and a NPDES permit to dispose of the wastewaters. Sludges from these tanks are subject to RCRA regulation when they are removed from the tanks. (Reauthorized in 1986.)

closure costs Funds used to close a site after landfilling has been completed. In addition to leveling and grading, closure costs include seeding and planting trees and shrubs, as well as post-closure monitoring.

composting The natural biological composition of organic material in the presence of air to form a humuslike material. Controlled methods of composting include mechanical mixing and aerating, ventilating the materials by dropping them through a vertical series of aerated chambers, or placing the compost piles out in the open air and mixing or turning them periodically.

Comprehensive Environmental Responsibility, Compensation and Liability Act; also Superfund (CERCLA) A federal law authorizing identification and remediation of abandoned hazardous waste sites. (Enacted in 1980, reauthorized in 1986.)

contract laboratories Laboratories under contract to the EPA's Contract Laboratory Program (CLP) that analyze samples taken from wastes, soil, air, and water or that carry out research projects.

designer bugs A popular term for microbes developed through biotechnology that can degrade specific toxic chemicals at their source in toxic waste dumps or in groundwater.

DDT The first chlorinated hydrocarbon insecticide. It has a half-life of 15 years and can collect in fatty tissues of certain animals. The EPA banned registration and interstate sale of DDT for virtually all but emergency uses in the United States in 1972 because of its persistence in the environment and accumulation in the food chain.

dredging Removal of mud from the bottom of water bodies using a scooping machine. This disturbs the ecosystem and causes silting that can kill aquatic life. Dredging of contaminated muds can expose aquatic life to heavy metals and other toxins.

electrostatic precipitator (ESP) An air pollution control device that removes particles from a gas stream. The ESP imparts an electrical charge to the particles, causing them to be attracted to metal plates or tubes inside the precipitator. Rapping on the plates causes the particles to fall into a hopper for disposal.

enforcement decision document (EDD) A document that explains to the public the EPA's selection of the cleanup alternative at enforcement sites on the National Priorities List. Similar to record of decision.

environmental audit 1. An independent assessment of the current status of a party's compliance with applicable environmental requirements. 2. An independent evaluation of a party's environmental compliance policies, practices, and controls.

extraction procedure toxicity test Test used to determine the toxicity characteristic of waste.

extremely hazardous substances Any of the 406 chemicals identified by the EPA on the basis of toxicity and listed under SARA Title III. The list is subject to revision.

feasibility study 1. Analysis of the practicality of a proposal; e.g., a description and analysis of the potential cleanup alternatives for a site or alternatives for a site on the National Priorities List. The feasibility study usually recommends selecting a cost-effective alternative and usually starts as soon as the remedial investigation is underway. Together they are commonly referred to as the RI/FS (Remedial Information/Feasibility Study). The term can apply to a variety of proposed corrective or regulatory actions. 2. In research, a small-scale investigation of a problem to ascertain whether or not a proposed research approach is likely to provide useful data.

filtration A treatment process, under the control of qualified operators, for removing solid (particulate) matter from water by passing the water through porous media such as sand or a man-made filter. The process is often used to remove particles that contain pathogenic organisms.

fluorocarbon Any of a number of organic compounds analogous to hydrocarbons in which one or more hydrogen atoms are replaced by fluorine. Once used in the United States as a propellant in aerosols, they are now primarily used in coolants and in some industrial processes. Fluorocarbons containing chlorine are called chlorofluorocarbons (CFCs). They are believed to be modifying the ozone layer in the stratosphere, thereby allowing more harmful solar radiation to reach the earth's surface.

fly ash Noncombustible residual particles from the combustion process carried by flue gas.

formaldehyde A gas used in the manufacture of particleboard and some plywood, especially for drawer fronts, cabinet doors, furniture tops, sub-flooring, some furniture, and mobile home decking. Low-level release, particularly under humid and warm conditions, can lead to breathing problems, eye and skin irritation, and dizziness. High concentrates have proven to cause cancer in test animals and probably humans. Pressed wood products should be used primarily outdoors.

Greenhouse effect The warming of the earth's atmosphere caused by a buildup of carbon dioxide or other trace gases. Many scientists believe that this buildup allows light from the sun's rays to heat the earth but prevents a counterbalancing loss of heat.

groundwater The supply of fresh water beneath the earth's surface (usually in aquifers) that is often used for supplying wells and springs. Because about 50 percent of groundwater is a major source of drinking water, there is growing concern over areas where leaching, urban runoff, agricultural or industrial pollutants, landfills, waste piles, and injection wells or substances from leaking underground storage tanks are contaminating groundwater. Groundwater is also used for about 80 percent of rural, domestic, and livestock needs, 40 percent of irrigation needs, and 25 percent of industrial needs.

Hazardous Ranking System The principal screening tool used by the EPA to evaluate risks to public health and the environment associated with abandoned or uncontrolled hazardous waste sites. The system calculates a score based on the potential of hazardous waste substances spreading from the site through the air, surface water, or groundwater and on other factors such as nearby population. This score is the primary factor in deciding if the site should be on the National Priorities List, and if so, what ranking it should have compared to other sites on the list.

Hazardous and Solid Waste Amendments (HSWA) Amendments to RCRA establish a timetable for landfill bans and more stringent requirements for underground storage tanks (1984).

hazardous substance 1. Any material that poses a threat to human health or the environment. Typical hazardous substances are toxic, corrosive, ignitable, explosive, or chemically reactive. 2. Any substance designated by the EPA to be reported if a designated quantity of the substance is spilled in the waters of the United States or if otherwise emitted into the environment.

hazardous waste By-products of society that can pose a substantial or potential hazard to human health or the environment when improperly managed. Possesses at least one of the four characteristics of ignitability, corrosivity, reactivity or toxicity, or appears on special EPA lists.

high-density polyethylene (HDPE) A material that produces toxic fumes when burned. Used to make plastic bottles and other products; also used as a liner material under landfills, pits, ponds, and lagoons.

high-level radioactive waste Waste generated in the fuel of a nuclear reactor, found at nuclear reactors, or nuclear fuel reprocessing plants. It is a serious threat to anyone who comes near the wastes without shielding.

humus Decomposed organic material.

hydrogeology The geology of groundwater, with particular emphasis on the chemistry and movement of water.

hydrology The science of dealing with water properties, distribution, and circulation.

incineration 1. Burning of certain types of solid, liquid, or gaseous materials. 2. A treatment technology involving destruction of waste by controlled burning at high temperatures; e.g., burning sludge to remove the water and reduce the remaining residues to a safe, nonburnable ash, which can be disposed of safely on land, in some waters, or in underground locations.

infiltration 1. The penetration of water through the ground surface into subsurface soil or the penetration of water from the soil into a sewer or other pipe through defective joints, connections, or manhole walls. 2. A land application technique where large volumes of wastewater are applied to land, allowed to penetrate the surface, and percolate through the underlying soil.

injection well A well in which fluids are injected for purposes such as waste disposal, improving the recovery of crude oil, or solution mining.

land application unit Discharge of wastewater onto the ground for treatment or reuse.

land bans Regulations that ban the disposal of hazardous waste in landfills without prior treatment. These regulations are being implemented in three phases, commonly referred to as first third, second third, and third third.

landfills 1. Sanitary landfills are land disposal sites for nonhazardous solid wastes. The waste is spread in layers, compacted to the smallest practical volume, and covered with material applied at the end of each operating day. 2. Secure chemical landfills are disposal sites for hazardous waste. They are selected and designed to minimize the chance of hazardous substances being released into the environment.

leachate A liquid that results from water collecting contaminants as it trickles through wastes, agricultural pesticides, or fertilizers. Leaching may occur in farming areas, feedlots, and landfills, and may result in hazardous substances entering surface water, groundwater, and soil.

liner 1. A relatively impermeable barrier designed to prevent leachate from leaking from a landfill. Liner materials include plastic and dense clay. 2. An insert or sleeve for sewer pipes to prevent leakage or infiltration.

listed waste Waste listed as hazardous under RCRA but that has not been subjected to the Toxic Characteristic Listing Process because the dangers they represent are considered self-evident.

low-level radioactive waste Wastes less hazardous than what is generated by a nuclear reactor, usually generated by hospitals, research laboratories, and certain industries. The Department of Energy, the Nuclear Regulatory Commission, and the EPA share responsibilities for managing them.

mass burn systems Systems that sift garbage to remove nonburnables, such as batteries and paint cans, which are then landfilled. The remaining material is loaded into an incinerator; the remaining ash residue is then landfilled.

maximum contaminant level The maximum permissible level of contaminant in water delivered to any user of a public water system. MCLs are enforceable standards.

methane (CH_4) A colorless, odorless gaseous hydrocarbon produced by marshes, rice fields, cattle droppings, and landfills. It is estimated that it contributes to 18 percent of the greenhouse effect.

National Emissions Standards for Hazardous Air Pollutants (NESHAPS) Emissions standards set by the EPA for an air pollutant not covered by National Ambient Air Quality Standards (NAAQS) that may cause an increase in deaths or serious, irreversible, or incapacitating illnesses. Primary standards are designed to protect human health, secondary standards to protect public welfare.

National Priorities List (NPL) Official list of hazardous waste sites to be addressed by CERCLA/SARA.

NIMBY "Not In My Backyard" syndrome that has come to characterize communities who protest and try to sabotage impending waste facilities.

nitrous oxides (N_2O) A colorless, nonvolatile, sweet-smelling gas also known in dental offices as "laughing gas." It is formed by burning fossil fuels, by microbes, and by the decomposition of chemical fertilizers. It is estimated to be responsible for 10 percent of the greenhouse effect.

off-site facility A hazardous waste treatment, storage, and disposal area that is located at a place away from the generating site.

on-site facility A hazardous waste treatment, storage, and disposal area that is located on the generating site.

operation and maintenance 1. Activities conducted at a site after a Superfund site action is completed to ensure that the action is effective and operating properly. 2. Actions taken after construction to ensure that facilities constructed to treat wastewater will be properly operated, maintained, and managed to achieve efficiency levels and prescribed effluent limitations in an optimum manner.

ozone A natural form of oxygen that serves as a protective layer shielding the earth from harmful ultraviolet radiation. Ozone can seriously affect the human respiratory system and is one of the most prevalent and widespread of all the criteria pollutants for which the Clean Air Act required the EPA to set standards.

paradichlorobenzene A familiar key ingredient in mothproofing agents and deodorizers. This compound has been found to be cancer-causing. Alternate products are cedar chips or wood for mothproofing and baking soda for deodorizing.

part A permit The first part of the RCRA permit application process for existing treatment, storage, and disposal facilities. Approved part A permits allow operation of a facility in interim status until part B permit approval is attained.

part B permit The second narrative section submitted by generators in the RCRA-permitting process. It details the procedures followed at a facility to protect human health and the environment.

pathogenic Capable of causing disease.

pathogens Microorganisms that can cause disease in other organisms or in humans, animals, and plants. They can be bacteria or viruses and are found in sewage, in runoff from animal farms, in rural areas populated with domestic and wild animals, and in water used for swimming.

pathology The study of the nature of diseases and especially of the structural and functional changes produced by them.

PCBs A group of toxic, persistent chemicals used in transformers and capacitors for insulating purposes and in gas pipeline systems as a lubricant. Further sale of new use was banned by law in 1979.

perchloroethylene A solvent used by most dry cleaners that has been identified as a cancer-causing agent in animals.

polyvinyl chloride (PVC) A tough, environmentally indestructible plastic that releases hydrochloric acid when burned.

post-closure The time period following the shutdown of a waste management facility or a manufacturing facility. For monitoring purposes, this is often considered to be 30 years.

potentially responsible party (PRP) An individual or company—including owners, operators, transporters, or generators—potentially responsible for, or contributing to, contamination problems at a Superfund site. Whenever possible, the EPA requires PRPs, through administrative and legal actions, to clean up hazardous waste sites they have contaminated.

precipitate A solid that separates from a solution of some chemical or physical change.

precipitation Removal of solids from liquid waste so that the hazardous solid portion can be disposed of safely; removal of particles from airborne emissions.

precipitators Air pollution control devices that collect particles from an emission.

radon A colorless, naturally occurring, radioactive, inert gas formed by decay of radium atoms in substrata soil or rocks. Next to smoking, it is believed to be the second leading cause of lung cancer. Radon can be reduced by sealing cracks in basement floors, ventilating crawl spaces, and having good air exchange throughout the house. While self-test kits are available at hardware stores, the EPA suggests professional help in installing radon-reduction measures.

record of decision (ROD) A public document that explains which cleanup alternative(s) will be used at National Priorities List sites where CERCLA trust funds pay for the cleanup.

refuse-derived fuel systems (RDF) Systems that process waste and burn only a fraction of the total material. Most involve shredding the waste and separating the material into light and heavy fractions. The light fraction contains paper and other easily burned items and is used as fuel for the incinerator.

remedial action The actual construction or implementation phase of a Superfund site cleanup that follows remedial design.

remedial design A phase of remedial action that follows the remedial investigation/feasibility study and includes the development of engineering and specifications for cleanup.

remedial investigation An in-depth study to determine the nature and extent of contamination at a Superfund site; to establish criteria for cleaning up the site; to identify preliminary alternatives to remedial actions; and to support the technical and cost analysis of the alternatives. The remedial investigation is usually done with the feasibility study. Together, they are usually referred to as RI/FS (Remedial Investigation/Feasibility Study).

remedial response A long-term action that stops, or substantially reduces, a hazardous substance, or the threat of a release of a hazardous substance that is serious, but not an immediate public threat.

residual Pollutants remaining in the environment after a natural or technological process has taken place: e.g., the sludge remaining after initial wastewater treatment or the particulates remaining in the air after the air passes through a scrubbing process.

response action A CERCLA-authorized action involving either a short-term removal action or a long-term removal response that might include, but not be limited to, removing hazardous materials from a site to an EPA-approved hazardous waste facility for treatment, containment, or destruction; containing the waste safely on-site; and identifying and removing the source of groundwater contamination and halting further migration of contaminants.

Resource Conservation and Recovery Act (RCRA) Regulates management and disposal of hazardous materials and wastes being generated, treated, stored, disposed, or distributed.

risk assessment The qualitative and quantitative evaluation performance that defines the risk posed to human health or the environment by the presence, potential presence, or use of specific pollutants.

risk management The process of evaluating alternative regulatory and nonregulatory responses to risk and selecting them. The selection process necessarily requires the consideration of legal, economic, and social factors.

rotary kiln incinerator An incinerator most commonly used for disposing of hazardous waste. It consists of a large rotating kiln, followed typically by an afterburner and scrubber.

scrubber An air pollution control device that uses a spray of water, reactant, or a dry process to trap acid pollutants in gaseous emissions.

site inspection The collection of information from a Superfund site to determine the extent and severity of hazards posed by the site. It follows, and is more extensive than, preliminary assessment. The purpose is to gather information necessary to score the site using the Hazard Ranking System and to determine if the site presents an immediate threat that requires prompt removal action.

sludge A semisolid residue produced from any of a number of air or water treatment processes of chemical and biological contaminants.

solid waste A nonliquid, nonsoluble material ranging from municipal garbage to industrial wastes that contain complex and sometimes hazardous substances. Solid wastes also include sludge, agricultural refuse, demolition waste, and mining residues. Technically, solid waste also refers to liquids and gases in containers.

Subtitle D Regulation Framework established under RCRA for federal, state, and local governments to manage solid waste, which includes voluntary implementation of solid waste management plans combined with minimum technical standards established by the EPA for new and existing solid waste management facilities. Solid wastes regulated by Subtitle D include: municipal solid waste, industrial and commercial nonhazardous waste, sludges from water supply and wastewater treatment plants, and special wastes, such as mining wastes, oil and gas waste, and infectious waste.

Superfund The program operated under the legislative authority of CERCLA and SARA that funds and carries out the EPA's solid waste emergency removal remedial activities. These activities include establishing the National Priorities List, investigating sites for inclusion on the list, determining their priority level on the list, and conducting and/or supervising the ultimately determined cleanup and other remedial actions.

Superfund Amendments and Reauthorization Act (SARA) A federal law reauthorizing and expanding the jurisdiction of CERCLA (1986).

surface impoundment Treatment, storage, or disposal of liquid hazardous wastes in ponds.

synthetic organic chemicals (SOCs) Man-made organic chemicals. Some SOCs are volatile, others tend to stay dissolved in water rather than evaporate out of it.

technology-based standards Effluent limitations applicable to direct and indirect sources, which are developed on a category-by-category basis using statutory factors, not including water quality effects.

toxic characteristics leaching procedure A proposed expanded version of the extraction procedure toxicity test. It broadens the universe of hazardous waste by including more organic compounds and metals at lower limits.

toxicology The science and study of poisons control.

Toxic Substances Control Act (TSCA) A 1976 federal law authorizing the EPA to gather information on chemical risks. Also regulates the production and distribution of new chemicals and governs the manufacture, processing, distribution, and use of existing chemicals. Among the chemicals controlled by TSCA regulations are: PCBs, chlorofluorocarbons, and asbestos. In specific cases, there is an interface with RCRA regulations; e.g., PCB disposal is generally regulated by TSCA, but hazardous wastes mixed with PCBs are regulated under RCRA.

treatment, storage, and disposal facility (TSD) Site where a hazardous substance is treated, stored, or disposed. TSD facilities are regulated by the EPA and states under RCRA.

trichloroethylene (TCE) A stable, low-boiling colorless liquid, toxic by inhalation. TCE is used as a solvent, metal degreasing agent, and in other industrial applications.

underground storage tank (UST) A tank with 10 percent or more of its volume underground and piping connected to the tank, regulated by RCRA. It is used to store petroleum products and hazardous chemicals regulated by CERCLA.

vinyl chloride A chemical compound used in producing some plastics and is believed to be carcinogenic.

volatile organic compound (VOC) Any organic compound that participates in atmosphere photochemical reactions except for those designated by the EPA administrator as having negligible photochemical reactivity.

volatile synthetic organic chemicals Chemicals that tend to volatilize or evaporate from water.

waste-to-energy A disposal method that uses high-power furnaces to incinerate solid waste. The heat given off is then sold as steam or electricity. With more than 100 plants in the United States, this method of disposal is quickly becoming an alternative to landfills. Most systems used are designed for incinerating municipal solid waste and are not applicable to some industrial solid waste streams. Systems that incinerate municipal solid waste can be broadly segregated into two categories: mass burn systems, which burn the entire stream of waste and refuse-derived fuel systems that process the solid waste first and then burn only certain fractions of the material.

waste pile A non-containerized mass of solid, non-flowing waste material that may or may not be enclosed by a fence, cover, or some other structure. Waste piles can be used for treatment or storage, but are also being outlawed in many states.

This glossary was compiled from information provided by the U.S. Environmental Protection Agency's *Glossary of Environmental Terms and Acronym List*, December 1989.

Bibliography

American Recycling Market Directory. Gale Directories, Detroit; 1989.

BioCycle. *The Biocycle Guide to Collecting, Processing and Marketing Recyclables.* BioCycle, Emmaus, PA, 1990.

Bookchin, Murray. *Remaking Society: Pathways to a Green Future.* South End Press, 1990.

Botkin, Daniel B. *Discordant Harmonies: A New Ecology for the Twenty-First Century.* Oxford University Press, 1989.

Brown, Lester and contributors. *State of the World: A Worldwatch Institute Report on Progress Toward a Sustainable Society.* Worldwatch Institute, Washington, D.C., 1990.

Caplan, Ruth and staff of Environmental Action. *Our Earth, Ourselves: The Action-Oriented Guide to Help You Protect and Preserve Our Environment.* Bantam Books, 1990.

Commoner, Barry. *Making Peace with the Planet.* Pantheon, 1989.

Council for an Energy-Efficient Economy. *The Most Energy-Efficient Appliances.* Council for an Energy-Efficient Economy, Washington, D.C., 1990.

Council of Governments. *Directory of Local Governments Recycling Practices.* Washington, D.C.

Council of State Governments. *Resource Guide to State Environmental Management.* Council of State Governments, Lexington, KY, 1988.

Council on Economic Priorities. *Shopping for a Better World: A Guide to Socially Responsible Supermarket Shopping.* Council on Economic Priorities, New York, NY, 1990.

Crampton, Norm. *Complete Trash: The Best Way to Get Rid of Practically Everything Around the House.* 1990.

Dodd, Debra Lynn. *Nontoxic & Natural: How to Avoid Dangerous and Everyday Products and Buy or Make Safe Ones.* Jeremy P. Tarcher, 1984.

Dodd, Debra Lynn. *Buyer's Guide to Water Purification Devices.* Guaranty Press, The Earth Works Group. *50 Simple Things You Can Do to Save the Earth.* The Earthworks Press, 1989.

Elkington, John and contributers. *The Green Consumer: Products You Can Buy that Don't Cost the Earth.* Barnes & Noble, 1990.

Environmental Law Institute. *Law of Environmental Protection.* Clark-Boardman, 1989.

Environmental Law Books. Clark-Boardman.

Environmental Economics. *Directory of Environmental Investing.* Environmental Economics, Philadelphia, PA, 1990.

Erickson, Brad, edited by. *Call to Action: Handbook for Ecology, Peace and Justice.* Sierra Club Press, 1989.

Gale Directories. *American Recycling Market Directory.* Gale Research Co., 1989.

Global Tomorrow Coalition, edited by Walter H. Corson. *The Global Ecology Handbook: What You Can Do About the Environmental Crisis.* Beacon Press, 1990.

Gribbin, John. *Hothouse Earth: The Greenhouse Effect and Gaia.* Grove Weidenfeld Press, 1989.

Head, Suzanne and Heinzman, Robert, edited by. *Lessons of the Rain Forest.* Sierra Club Books, 1989.

Heloise: Hints for a Healthy Planet. Putman's, 1989.

Hines, Patricia H. *Earth Right: Every Citizen's Guide.* Prima Publishing, 1989.

Lamb, Marjorie. *2 Minutes a Day for a Greener Planet.* Harper & Row, 1989.

Lyman, Francesca. *The Greenhouse Trap.* Beacon Press, 1989.

MacEachern, Diane. *Save Our Planet: 750 Everyday Ways You Can Help Clean Up the Earth.* Dell Publishing, 1990.

Machlin, Jennifer L. and Young, Tomme R. *Managing Environmental Risk: Real Estate and Business Transactions.* Clark-Boardman, 1989.

Managing Planet Earth: Readings from *Scientific American Magazine.* W.H. Freeman & Co., 1990.

Naar, Jon. *Design for a Livable Planet.* Harper & Row/Perennial Library, 1990.

Oppenheimer, Michael and Boyle, Robert H. *Dead Heat: The Race Against the Greenhouse Effect.* Basic Books, 1989.

Orloff, Neil and Sakai, Susan. *Community Right-to-Know Handbook: A Guide to Compliance with EPCRA.* Clark-Boardman, 1988.

Pearson, David, with Dadd, Debra Lynn. *The Natural House Book.* Simon & Schuster, 1989.

Rifkin, Jeremy, edited by. *The Green Lifestyle Handbook: 1001 Ways You Can Heal the Earth.* Henry Holt & Co., 1989.

Selmi, Daniel P. and Manaster, Kenneth A. *State Environmental Law.* Clark-Boardman, 1990.

Shopping for a Better World: A Guide to Socially Responsible Supermarket Shopping. 1800 Popular Brands rated from A to Z. Council on Economic Priorities, 1990.

Sombke, Laurence. *The Solution to Pollution: 101 Things You Can Do to Clean Up Your Environment.* MasterMedia, 1989.

Steger, Will and Bowermaster, Jon. *Saving the Earth: A Citizen's Guide to Environmental Action.* Knopf, 1989.

Tarlock, A. Dan. *Law of Water Rights Resources.* Clark-Boardman, 1990.

Want, William. *Law of Wetlands Regulations.* Clark-Boardman, 1990.

White House Council on Environmental Quality Report 1990. Council on Environmental Quality, 1990.

World Environmental Directory, 5th Edition. Business Publishers, Inc., Silver Spring, MD, 1989.

Magazines

BioCycle, Box 351, Emmaus, PA, 18049, $55.

Design Spirit, 438 Third Street, Brooklyn, NY 11215, $16 a year.

Discover, P.O. Box 359087, Palm Coast, FL 32035-9944, $26.95.

Environment, Heldref Publication, 4000 Albemarle Street, NW, Washington, D.C., 20077-5010, $23.

Garbage, The Practical Journal for the Environment, P.O. Box 51647, Boulder, CO 80321-1647, $21.

In Business, Box 323, Emmaus, PA 18049, $18.

Journal of Waste Recycling, BioCycle, Box 351, Emmaus, PA 18049, $55 a year.

Other Magazines that Feature Articles on the Environment

Entrepreneur, The Small Business Authority, P.O. Box 19787, Irvine, CA 92713-9438, $3 an issue.

Harrowsmith, America's Intelligent New Magazine, Ferry Road, P.O. Box 1000, Charlotte, VT 05445-9984, $14.97.

INC., The Magazine for Growing Companies, P.O. Box 51534, Boulder, CO 80321-1534, $25.

Mother Earth News, The Original Country Magazine, P.O. Box 3122, Harlan, IA 51593-2188, $13.95.

Scientific American, 415 Madison Avenue, New York, NY 10017, $24.

SUCCESS, The Magazine for Today's Entrepreneurial Mind, P.O. Box 3036, Harlan, IA 51593-2097, $11.97.

The Environmental Forum, The National Journal of Environmental Affairs, Environmental Law Institute, 1616 P Street, NW, Washington, D.C., 20036, $60.

Index

A

B

S

T